T0094058

Disasters

Dr. Asim K. Dasgupta

First edition 2008
(published by Athena press, London, UK)

Second edition (updated and extended) 2011
(published by authorHOUSE UK Ltd.)

AuthorHouse™ UK Ltd.
500 Avebury Boulevard
Central Milton Keynes, MK9 2BE
www.authorhouse.co.uk
Phone: 08001974150

Second edition published by AuthorHouse January 2011

ISBN: 978-1-4520-6584-7 (sc)
ISBN: 978-1-4520-6585-4 (e)

This book is printed on acid-free paper.

About the Author

DR ASIM K DASGUPTA is a medical doctor and a former Occupational Health Consultant in the National Health Service, United Kingdom. He also worked in private and public sector industries. His main interest is in Rehabilitation and Environmental Medicine which involved him in many investigations and research. He worked and travelled to many world disaster zones and now he engages himself in researching, travelling, writing, and publishing of this fascinating subject.

This book is dedicated to the people who lost their lives, or became disabled or orphaned, as a result of various disasters across the world.

A Note on Place Names

In recent years, the names of several countries and cities in Asia have been changed. For example, Bombay is now called Mumbai, Burma is Myanmar, Ceylon is Sri Lanka, Calcutta is Kolkata, Madras is Chennai and Peking is Beijing. The places mentioned in this book are mostly the new names. The most recent change was of Calcutta and according to the relevant time period, the author has used both 'Calcutta' and 'Kolkata'.

Contents

Preface for 2nd edition

This is the new edition of my earlier book entitled 'Disasters – A wander down memory Lane' which was originally published in the year 2008 by the Athena Press, London, UK.

Since the first edition of the book was published, natural disasters have continued to occur with great impact on the global environment, often resulting in human catastrophe. My journey covers both near miss incidents and some of the world's greatest disasters. Furthermore, man-made climate change in the form of global warming is adding to the misery of the human population. Many health professionals think that climate change is now the biggest threat to the well being of the twenty -first century. If we do not act now, the survival of the global community will be called into question. With the support and encouragement of many readers, I have written this new edition with the latest information available to me. A new chapter on 'Global Climate Change' has been added to this edition. In addition, new information is described on the incidents in the following chapters:

1. Earthquakes and Tsunamis
2. Volcanoes
3. Floods and Dams
4. Storms, Tidal Waves and Floods
5. Gales, Snow and Hail
6. Droughts, Famine and Starvation
7. Accidents and Accidental Disaster

The chapter 'The Sun and Solar Eclipses' has been retitled as 'The Sun, the Solar System and Solar Eclipses' with the information on the solar system and the space shuttle disasters has now been incorporated.

The chapter Migration, Refugee and Asylum Seekers is also now changed to Human Movement.

Many photographs are new and have replaced the first edition photographs. The sources of information are all the same as mentioned in the 'Preface' of the first edition.

Preface for 1st edition

Since my childhood, I have been curious about the sudden happenings of nature, like lightning, solar eclipses, thunderstorms, cyclones and tornadoes. Over the years, I have come across natural calamities such as drought, floods, mudslides, landslides, earthquakes and volcanoes. As a medical doctor, I have often been involved in relief work and have visited and seen many scenes of death, injury, rehabilitation and deprivation. I felt the need to write this book, describing my journey from childhood to the present time and hence, I have drawn on more than fifty years of autobiographical perspective. My interest in Earth science and global warming, which is affecting the planet and our environment, has also been reflected.

This book consists of thirteen chapters with photographs. One chapter considers accidental disasters and the rest focus on natural and man-made disasters. In this book, I not only consider the calamity and its effects but also try to narrate the associated and underlying beauty and wonder of nature, famous historical monuments and World Heritage Sites, which might encourage people to visit those places or countries.

I have chosen the phrase 'down memory lane' for the title of my book because over fifty years' experience has allowed me to draw some conclusions that I hope may contribute to our human capacity to respond in the future to what ancient societies called 'ill-starred events'.

Modern science allows ever-growing possibilities for predicting catastrophe; scientists, engineers and administrators can now anticipate and plan resources and procedures to cope with the inevitable destruction and misery of disaster. As a physician, scientist and traveller, I have written this book and drawn conclusions on the role of technology, and the prevention and betterment of humankind. I trust my life experiences will contribute to reducing the impact of the many scenes of death, injury and deprivation due to disasters or calamities.

Information about places that I was unable to visit came from a variety of other sources, including journalists, relatives and friends. To write this book, I also depended upon newspaper articles, journals, travel and Earth science books, radio, television and internet search engines.

Chapter One
Earthquakes and Tsunamis

Earthquake-resistant structure, Sacsayhuman, Peru

According to the Oxford English Dictionary, an earthquake is defined as 'a convulsion of the superficial parts of the Earth due to the release of accumulated stress, as a result of faults in strata or volcanic action'. A tsunami is 'a long, high sea wave caused by underwater earthquakes or other disturbances, such as volcanic eruption or sediment slump'.

Although an earthquake is quite simply a shaking, it is no doubt the most terrifying event of the Earth. The violent movements of the Earth's surface occur suddenly, resulting in devastating effects on human lives and land mass.

In Hindu mythology, it is said that the god Brahma is holding the Earth with one hand and, when he gets tired, changes his hand. During the short spell of change of hands, the Earth rocks, resulting in devastating effects.

In ancient Greece, people used to worship the god Poseidon, the 'earth-shaker', and believed that he controlled the earthquake.

A recorded history of earthquakes goes back to 2300 BC China and, since then, history has witnessed many such catastrophes. If one considers the record of the most destructive known earthquakes the world has seen, China has had the most. So I was very interested in visiting China. In spring 2004, I had that opportunity.

Not far from Beijing, there is a pyramid-shaped memorial in the city of Tianjin for the victims of the 1976 earthquake, when 255,000 people were killed (although the unofficial toll was 655,000). This was the result of an earthquake that struck the north-east of China on 27 July with a magnitude of 7.5 on the Richter scale.

The effects of an earthquake depend upon its magnitude and intensity. Intensity is the degree of shakiness caused by earth quake, and magnitude is a measure of earthquake, calculated by the Richter Scale. The Richter Scale was devised by the American seismologist Charles Richter in 1935. The intensity depends upon the magnitude; a magnitude below three is so minute that it usually passes virtually unnoticed, but from three to five, it can be felt. Above five, the damaging effects start, and above seven, the Earth reveals its destructive power. The maximum intensity is at the earthquake epicentre and then decreases with distance from the epicentre. The severity of an earthquake can be described on the basis of magnitude (Table 1).

Table 1: Richter Scale Severity

Magnitude	Effects
0–2	Not felt
2–4	Minor
4–5	Small
5–6	Moderate
6–7	Strong
7–8	Major
8+	Great

With modern scientific instruments (seismographs) and improved communication skills, the earthquake zones of the world can be easily mapped. In particular, now that the calculation of epicentres is possible, standardised mapping is widely available. It is possible to distinguish from seismographs whether an earthquake resulted from a bomb or from the Earth itself.

China pioneered the primitive version of the modern seismograph and a Chinese man, named Han Sin, invented it in the second century BC.

In modern China, I was surprised to see so many tall, newly constructed buildings, especially in Beijing and Shanghai. Many more skyscrapers were under construction. There is no doubt that China is booming economically. However, I was wondering how these tall buildings would stand against the destructive forces of earthquakes, as China is situated in an earthquake zone. When I tried to discuss the effect of earthquakes on China's historical buildings and the present-day skyscrapers, our guide pointed out that some of the oldest monuments and buildings in the Forbidden City had survived earthquakes over the centuries.

However, this I could not support fully when I visited and saw the cumulative effects of earthquakes on an ancient Buddhist pagoda in a place called Xi'an, the capital of China's Shaanxi Province. The top two stories of the fifteen-storied Little Goose Pagoda, a Buddhist temple built in AD 707, collapsed in the 1556 earthquake, and I also noticed some cracks on the walls of the pagoda.

Apart from damage to historical buildings, this earthquake took the lives of 830,000 people and its magnitude was 8.0 on the Richter Scale. It happened on 23 January 1556.

The world's most durable earthquake-resistant construction is found in the fortress and temple of Sacsayhuman in Cusco Valley in Peru. Cusco, the capital of the ancient Incan empire, lies in the Andes at an altitude of 3,520 m. The stone structures were built by the Incas in such a way that they are locked and dove-tailed into position, making them earthquake-proof. Over the years, many devastating earthquakes have shaken the Andes but the blocks are still in place, while the Spanish cathedral in Cusco has collapsed a number of times.

My wife, Supta and I visited Peru in July 2007, nearly three weeks prior to a powerful earthquake that killed 510 and injured 1,500 people. The magnitude was 8.0 on the Richter Scale, and the epicentre was in the Pacific Ocean, ninety miles (145 kilometres) south-east of the capital, Lima. The most affected town was Chincha Alta. This earthquake occurred on 15 August 2007, and was followed by several aftershocks. A tsunami generated by the earthquake affected Peru, Chile, Ecuador and Colombia, but the effects were less catastrophic than one expected.

Due to the nature of natural catastrophes, it is very difficult to give exact numbers of earthquake victims in various parts of the world. However, in the last century, earthquakes have probably claimed the lives of one and a half to two million people throughout the world.

I remember my childhood experience of an earthquake in Calcutta, now known as Kolkata, India. It was in 1950 and we were having a meal at our home. Suddenly, we felt our plates start to move to one side and it seemed that the building was tilting.

My father started to shout, 'Come out of the building, come out of the building!' I could hear the sounds of the bell, the gong and the blow of the conch shell from nearby. I was puzzled by the sounds, as to whether they were to alert the people or whether they were some kind of worship to the Hindu gods to prevent disaster. The earthquake lasted a few seconds and its magnitude must have been below five, so it did not cause any damage. It was an exciting time; we all stayed out on the street in front of our house. We were told not to go inside as there was a chance of recurrence, or aftershocks. I cannot recall how long we waited before we were allowed to go inside. I was probably not scared because, as a child, I had no idea of how devastating the effects of an earthquake could be. I had mixed feelings of curiosity and fun.

The news of this earthquake made headlines in the following day's newspapers. It was the Assam earthquake; the epicentre was on the India–China border (the magnitude was 8.6–9.0 on the Richter Scale) and it then spread through upper and lower Assam, Darjeeling, Dhaka and Calcutta. A total area of 1,794,000 square kilometres was affected, of which 49,700 square kilometres suffered a more severe disaster, in which 1,530 people were killed.

The largest earthquake in Calcutta took place in 1737, in which 3,000,000 people died. However, it is now believed that the high fatalities were due to storms and floods, rather than the earthquake itself. This was the result of a cyclone which struck Calcutta and the surrounding areas at the same time (7 October 1737).

The Indian subcontinent is always vulnerable to earthquakes, being situated in the Asian collision zone, where continuing pressure from one plate to another under the Himalayas generates powerful and frequent earthquakes.

The rescue of people buried or trapped under the rubble is the main problem associated with earthquakes. International rescue and experts are available nowadays, and many countries offer help with their search and rescue teams. Specialised equipments and techniques, even search dogs, are promptly available. This help is needed while the chance of survival still persists, even if people have been trapped for many days since the incident occurred. It has been seen that people can survive for long periods without food and/or water and sometimes they have dug their own way out. Of course, this will depend upon how deeply someone is trapped or buried and the degree of injury they sustain. A prime example of this was in the 2003 Iran earthquake, where a man was found alive after being trapped for thirteen days. A similar example is found on 8 October 2005, in the Kashmir earthquake, when a twenty-year-old farm worker miraculously survived for twenty-seven days after being buried in the rubble of his house. The longest survival in earthquake disasters, most probably was in Quetta earthquake in the Indian sub-continents, which took place on 31st May 1935, when a merchant was buried in the debris of his shop but was able to escape after 47 days.

In August 1999 there was a great earthquake in Izmit, Turkey, with a magnitude of 7.4. People blamed substandard construction and bad soil structure for the collapse of the buildings that killed thousands of people: 20,000 people died. The aftershock impact was great, both economically and psychologically. When I visited Turkey three years after the Izmit earthquake, people were still living in temporary shelters and emergency housing. However, reconstruction was also ongoing and people were moving to newly constructed sites.

Psychologists and social workers working in this area said that the earthquake had changed people's lives and they had to deal every day with problematic family relationships, anxiety, sleeping disorders and alcohol abuse. Whatever lessons are learnt from this earthquake, the psychiatrists, psychologists and social workers' efforts at rehabilitation in this type of disaster are no doubt crucial.

Turkey lies on the collision zone. During the last century, there were twelve significant earthquakes and one catastrophic one. In the century, the most recent one was 8thMarch 2010 with the magnitude of 6.0 (Richter Scale) which killed at least 57 people and more earthquakes are forecast for this century. So, in order to prevent high mortality, it appears that Turkey should give more importance to building houses with solid structured foundations.

Like Turkey, India has serious earthquake problems, and a recent earthquake in January 2001, in the western Indian city of Bhuj, Gujarat, killed 18,602 and injured 165,229 people. Some of the victims were first-time visitors to the country. With aftershocks, the figure increased to 20,000 deaths and 167,000 injured. According to the government of India, 15.9 million people were victims of the earthquake and its aftershocks. Over one million homes were destroyed.

Unlike Turkey, large areas of countryside were affected, resulting in the destruction of crops, livestock and agriculture infrastructures. 20,618 cattle were killed.

UNICEF estimated that five million children under the age of fourteen were affected and some three million children lost family members. Children suffered from trauma and needed medical or psychosocial counselling, as there were problems with recovery.

Within two weeks of the earthquake, 2,232 people underwent major surgery. Amputation was not uncommon. The magnitude of the earthquake was 7.7 on the Richter Scale and the epicentre of the earthquake was sixty-nine kilometres north-east of the city of Bhuj.

A powerful earthquake of 6.7 in magnitude struck the ancient Silk Road city of Bam in south-eastern Iran at about 5.30 p.m. On 26 December 2003, killing 42,000 people and injuring 30,000. Many were homeless and traumatised. 60 per cent of the city was destroyed, collapsing buildings and damaging the 2,000-year-old Citadel, a historic mud-brick fortress and a UNESCO World Heritage Site. Two hospitals were completely destroyed and field hospitals were set up. There was an aftershock of 5.4 in magnitude about two hours later. Although the magnitude was low, the damage was tremendous and it was probably due to the type of houses and their construction, which were mostly made of mud-bricks. International rescue teams and forty-six countries offered help and assistance to rebuild shattered lives, even though Iran is politically isolated. Prince Charles came from Britain to visit the earthquake site; some people considered this a diplomatic move to improve the relationship between Iran and Britain while the Iraq war was ongoing. Earthquakes are not uncommon in Iran and the last major earthquake (of 7.3–7.7 in magnitude) was on 21 June 1990, when 50,000 people died.

The earthquake map shows that earthquakes affect the wealthier nations too, and the prime example of this is the 1906 San Francisco earthquake in the USA. When I visited San Francisco in 1988, our guide was narrating the story of the 1906 earthquake from inside a beautiful seventeenth-century church, noting how the building has withstood the various earthquakes. Later, I learned how buildings in San Francisco, constructed on solid rocks, withstood the devastating effects of the 1906 earthquake.

On the other hand, buildings with poor foundations collapsed easily and the devastation was greater. Solid-foundation buildings on solid rocks survived the earthquake, but secondary effects like fire destroyed those structures in the wealthier area of San Francisco in 1906. At the end of the tour, the guide told me that an earthquake had been predicted to take place in San Francisco on that very day. However, it did not occur on that day or that year, and I was rather disappointed that the prediction was wrong, or that the guide was telling a fib. However, later on I verified this and found that the 1988 Geological Survey of USA had forecast that this particular zone would have an earthquake in the next thirty years but could not predict exactly when it would happen. It did happen the next year (1989) on the Santa Cruz Mountain, about sixty-two miles (one hundred kilometres) south-east of San Francisco, injuring 3,757 with a death toll of sixty-two people and a great deal of damage to property, highways and bridges.

San Francisco stands on several earthquake fault lines and there are seven faults just in the Bay Area, so it is no wonder that San Francisco is vulnerable to earthquakes!

In 1989, its most damaged area was the landfill site where houses were built. Bay Bridge, which connects San Francisco, Treasure Island and Auckland, was affected and the most damaged part was the Auckland side. During my visit in 2006, I saw strengthening work taking place on the Bay Bridge, with the construction of an earthquake-proof structure.

In the same year, San Francisco celebrated the one hundredth anniversary of the 1906 earthquake that destroyed 75 per cent of the city. Fires blazed for four days. San Francisco is always prone to fire hazard.

In Britain, minor earthquakes occur from time to time, but the magnitudes of most of them are so low that they are hardly noticeable. The last small earthquake took place on 28 April 2007 at 8.18 a.m., measuring 4.3 on the Richter Scale, which struck about ten miles south-east of Folkestone, Kent. One person was injured and was taken to the local hospital with minor head and neck injuries. Some houses were damaged. According to the British Geological Survey, this earthquake resulted from a complex network of faults under the English Channel. Earlier records suggest that there have been earthquakes at the same location in the past. One was in 1382 and another was in 1580. In 1580, an earthquake of about 6.0 on the Richter Scale killed two people and some believe that there was a tsunami following this earthquake!

Many people do not know that earthquakes can cause tsunamis. Tsunamis are a series of sea waves caused by a massive shift of the seafloor during an earthquake. Tsunami means 'harbour wave'; the Japanese word 'tsu' means harbour, and 'nami' means wave. The name is given because of the way the waves crash into harbours. Tsunamis are most often seen in the Pacific Ocean and they mostly affect Japan's coastline. Japanese records on this subject suggest that there was a tsunami of two to three metres high along the eastern coastline of Japan as a result of an earthquake measuring 9.0 on the Richter Scale on the Pacific north-west on 26 January 1700 at 9 p.m. Japan is one of the most earthquake-prone countries of the world, as it is situated on top of four tectonic plates. Therefore the country has developed a sophisticated tsunami warning service run by Japan Meteorological Agency (JMA).

The tsunami that was triggered by an earthquake on 1 April 1946 in Alaska was the world's fastest-moving tsunami, travelling at more than 700 km/h to strike Big Island of Hawaii and killed more then 170 people. The most affected areas were Hilo, Laupahoehoe and Pololu Valley. Some waves were up to 55 ft in height. Many sugar cane plants and train tracks and bridges, used to transport this crop, were destroyed. When I was in Big Island in March 2008, besides Hilo, I also visited Laupahoehoe Point. Laupahoehoe means 'Leaf of Lava' and it was once a well-established little community, with a busy harbour and the location of Laupahoehoe Sugar Company. In April 1946, the three tidal waves of about 30 ft swept over the peninsula destroying sugar plants and the local school, and killing many residents. At present, there is a monument on the site of the school to commemorate the twenty-four victims who lost their lives.

Tsunamis are not uncommon on the island of Hawaii, and since the early 1800s about fifty tsunamis have been reported, two of which were major: one in 1946 and the other in 1960. On both these occasions Hilo was affected. The 1960 tsunami originated from Chile, and although its arrival time was predicted correctly, sixty-one lives were still lost. It was following the disastrous tsunami of 1946 that the tsunami warning system for the Pacific was developed by the USA and its headquarters in Honolulu. The tsunami warning centre receives information from seismometers and tide-gauge stations all around the Pacific Ocean. If there is an earthquake with a possible tsunami, then the warning centre puts out a 'tsunami watch', alerting civil defence and other authorities.

When the first positive evidence of a tsunami comes from the tide stations near the epicentre of the earthquake, Honolulu centre issues a 'tsunami warning' informing those at risk of the estimated arrival time for the first wave. Local authorities take action to evacuate the people at risk from low-lying areas. When I was travelling along the coastline on the island of Oahu, many beautiful beaches were empty and I noticed red flag signals which indicated high tide and rough seas.

Sugar plantations and the townships and railways associated with them used to dominate Big Island but are no longer in existence. The last sugar cane plantation closed down in 1996.

This was not due to tsunami but to the plantation's inability to compete in the global economy. Many people lost their jobs and found it hard to get other employment. Some committed suicide or became drug addicts or alcoholics. Many of the once-thriving plantation-based communities have become ghost towns.

In 2004, the day after Christmas, the tsunami waves in the Indian Ocean rocked the world; the power of the waves resulted in the worst catastrophe ever across South Asia.

On Boxing Day, 26 December 2004, at about 10 a.m., I received a phone call from my son-in-law, Rahul, saying that a big tsunami had hit southern Asia. Rahul Tandon works for BBC Radio Five Live, and at that time I did not take much notice.

However, I asked him, 'Are you going to South Asia to cover the disaster?' Sometimes he goes away to cover such events. He had been in India at the time of the Gujarat earthquake and reported on the incident for BBC Radio Five Live from the epicentre, near the city of Bhuj. This time, he replied, 'I do not know yet'. However, to my surprise, this time he did not go; instead, my daughter, Rumella, who is his wife and works for the BBC World Service, went. It was big and sad news throughout the world. The tsunami was the result of an undersea earthquake of 9.0 on the Richter Scale that struck at 12.59 a.m. On 26 December 2004 and the epicentre was sixty miles off the island of Sumatra, Indonesia. Next to Sumatra, the Andaman Islands in India was one of the first affected places, 860 kilometres north of the epicentre. Rumella and her colleague also went to the Andaman Islands. This was not Rumella's first visit to the Andaman. When she was ten years old, she visited with us. Naturally, when she returned from the disaster area, she showed me some photographs of the islands.

I asked her, 'Did you recall the areas which you had visited previously in your childhood? Did you notice any obvious damage by the tsunami of those areas?'

Rumella answered, 'Port Blair was not affected much, but Car Nicober and the Nicober group of islands were mainly affected. 901 people were killed and 5,914 more are missing, and many bodies were washed up on Port Blair beaches.'

She also said that, 'We were not allowed to go to many places, especially to the Nicober Islands.'

The latest official death toll from the tsunami in the Andaman and Nicober Islands reports that 1,316 were killed and 200 bodies were washed up on Port Blair beaches alone.

Port Blair is the state capital of the Andaman and Nicober groups of islands, which we visited in 1982. During my visit to Andaman, I was interested in the coral reefs and the ancient tribes. Although most of the islands of Andaman and Nicober were uninhabited, there are tribes of Mongoloid and Negrito stock that live near the coast and in the dense forests surrounded by the sea: they are the Jarawas, Great Adamanese, Onges, Sentinelese and Shopers. Most of these groups escaped the tsunami by moving to higher ground and into the deep forest. To find the Negrito tribes in this part of the world is unexpected, as anthropological and deoxyribonucleic acid (DNA) evidence suggests they originated in Africa. Did the Ice Age play a role? Some say that Negrito tribes migrated from Africa 70,000 years ago through Indonesia.

We saw plenty of beautiful, colourful living coral and coral reefs while walking in the shallow water near the Beach of Jolly Buoy Island. It was painful to the feet but beautiful to see, even with the naked eye.

Unlike the Maldives, the Andaman coral reefs were hardly affected by the 1998 worldwide coral bleaching, but Andaman coral did not escape the tsunami in some places. Coral reef damage is more noted in the Andaman, where the land level has risen above the present-day mean high tide. As a result of the 26 December 2004 earthquake and tsunami, some of the character of the Andaman Islands has been changed. There has been an uplift of coral reefs along the Western Andaman. Islands like Interview Island and North Sentinel rose by up to two metres. Several new sandbars and islands have emerged along the northwest coast of Middle Andaman Island. The east coast of South and Middle Andaman sank by up to two and a half to three metres, and the entire Andaman chain has tilted to the east-south-east. This remarkable information I got from Mike Searle, whom I went to see in Oxford. Mike Searle is a lecturer and senior research scientist in the Earth Sciences Department of Oxford University.

When I visited the Andaman Islands in January 2007, as I passed through Sipi Ghat, near Port Blair, I saw some of the paddy fields and coconut tree plantation areas that were fully submerged with seawater due to geological changes following the tsunami and earthquake. This I noted when I was going to Baratang Island along the Andaman's Trunk Road.

The Andaman Trunk Road runs through part of Jarawa country. The Jarawas live along the western coast of South and Middle Andaman. In 1982, when I was visiting the Andaman Islands, this road was under construction and there was no public access. The Jarawas are very hostile and some people have been killed by them. Since the construction of the Trunk Road, vehicles are allowed to travel through, accompanied by an armed guard. Our car also had an armed guard; we were told not to stop anywhere and the windows of our car were to be closed at all times while we were passing through the Jarawa-designated territory, as they are still occasionally shattered by Jarawa arrows, and people are sometimes killed. On the way, we saw a Jarawa mother and child sitting on the roadside, and the next day, on our way back, we saw three Jarawa men with their bows and arrows, crossing the road. We wanted to stop and take some photographs but were not allowed. I was with Pinaki Bandopadhyay, a geologist from the Geological Survey of India.

The gigantic wave that rose from the floor of the ocean crashed into the coasts of Indonesia at 1.30 a.m., Thailand at 3 a.m., southern India at 3.30 a.m., Sri Lanka at 4 a.m., the Maldives at 6.40 a.m., Malaysia at 7 a.m., and, after travelling 2,800 miles across the Indian Ocean, it hit the African country Somalia at around 4 p.m. It took fifteen minutes to hit Sumatra, thirty minutes to hit the Andaman Islands, ninety minutes to hit Thailand, two hours to hit Sri Lanka, three hours, thirty minutes to hit the Maldives and seven hours to reach Somalia. Mombasa, Kenya, and the Seychelles were also affected, twelve hours after the earthquake. It seems that the tsunami waves even reached 3,700 miles west of the epicentre. Altogether, fifteen nations were affected and there was a series of tsunamis that washed away cities, towns, villages and resorts, carrying away people, houses, hotels, fishing boats, cars, lorries, trains and buses.

Tourists were sunbathing and some had been swimming or riding jet skis, strolling or jogging on the sand, and children were playing on the beach. Fishermen had returned with their catch and were dragging their fishing boats up onto the seashore; women were lining up to buy fish to sell on the mainland. The sea, which was slightly restless, had built up and, within five to ten minutes, the wall of water struck. The killer waves, which in places were thirty feet high, flattened everything in sight and the sea became calm after sucking people in. Some bodies were washed ashore.

At the outset, nobody realised the extent of its devastation but, when the waves receded, the magnitude of the tragedy was revealed. Over 235,000 people were dead, over one and half million people were homeless and many people, including a number of tourists, were found to be missing. The full extent of deaths from the tsunami may never be known. Children were the most tragic victims, and one third of the dead appeared to be children. Survivors saw their loved ones drowned in front of their eyes. The people who survived each had a story to tell. Many had miraculous escapes. There were fears that diseases such as diarrhoea, typhoid and malaria might claim the lives of many of those who had survived. Children were orphaned or abandoned or adopted. Even a year after the tsunami, many parents believed that their missing boy or girl was still alive.

The devastating tsunami was triggered by an earthquake beneath the surface of the Indian Ocean, where the India plate slipped below the Burma plate and pressure built up. Sudden movement of the plates caused the earthquake, and seismologists were able to detect a tremor of a magnitude of 9.0 on the Richter Scale on 26 December 2004. Countries thousands of miles away from the epicentre could have been warned in time to evacuate coastal populations, but unfortunately there was no warning. Scientists at the Pacific Tsunami Warning Center (PTWC), based in Hawaii, issued a bulletin about the earthquake but did not mention the possible tsunami. Fifty minutes later, the PTWC raised the first estimate of the earthquake's power (8.5 in magnitude) and said that a local tsunami was possible. Some will argue that the magnitude of the earthquake was monitored, but not the tsunami. There is no tsunami warning system in the Indian Ocean, like the Pacific Tsunami Warning System. In the Pacific, sensors on the sea floor monitor underwater earthquakes. The sensors pass on the information to floating buoys on the surface whenever they detect any change in sea pressure. Information is then relayed to satellites that pass it on to the earth stations. The twenty-six member nations of the Pacific Tsunami Warning System are then notified of the approaching danger and people are alerted through the media.

This Indian Ocean catastrophe was the first time I saw how the world community could be united in sorrow. There was an overwhelming response from the world community to the United Nations' emergency appeal and it was the biggest ever relief effort by the UN. Up to seven billion pounds in donations were pledged around the world. Britain raised millions of pounds through the British Disasters Emergency Committee, representing twelve large charities. The British public responded overwhelmingly. I also donated an amount. There was a three- minute silence throughout the UK and Europe following the disaster, in which I also took part, as a mark of respect to the victims. I gather that £432 million has already been allocated to rebuild shattered lives in the disaster areas. Many people, including some doctors from the UK, especially doctors of Sri Lankan origin, went to work in disaster areas in response to the Sri Lankan government's appeal. I was not Sri Lankan and so I planned a private visit to one of the Indian Ocean's tsunami- affected regions. The country I chose was the Republic of Maldives, which I had never visited before. Some predict that some of the Maldivian islands might disappear into the sea by the end of this century if global warming continues at its current rate.

The Maldives consist of twenty-six coral reefs forming atolls. An atoll is a ring of coral reefs or coral islands, or both, surrounding a lagoon. A lagoon is colourless but from the surface appears blue, and the size of the coral island varies from one to one hundred acres. The Maldives are made up of such a chain of 1,190 low-lying coral islands situated south-west of India and Sri Lanka, extending up to the equator in the Indian Ocean. The chain is 744 kilometres (158 miles) long and 118 kilometres (73 miles) wide, and the elevation is not more than three metres above sea level. The waves, which rose from the ocean floor near the epicentre, reached the Maldives at approximately 9.20 a.m. On 26 December 2004. The waves were up to five metres high and the disaster caused damage to sixty-nine out of 199 inhabited islands. Fourteen islands were completely devastated and all of the people were evacuated. 20,500 were displaced from their homes, eight-one people were killed and twenty-seven people are still missing.

The tsunami damage in the Maldives was not, however, as devastating as in other Indian Ocean countries, although two thirds of the Maldives' population were affected. This is because of the unique geographical nature of the Maldives. Most islands escaped damage, but some of them and the atolls were hurt. The most affected area was the east coast, mainly North Male atoll, South Male atoll and Kandholhudeo Island in Raa Atoll in the north. All except eight homes were uninhabitable as a result of the tsunami, and the government is planning to move everyone to a new site twenty kilometres away, on an uninhabited island called Dhuvaafaru.

I was told that about 12,000 islanders were still living in temporary accommodation or in the homes of relatives on neighbouring islands.

Many resort islands were completely unaffected by the tsunami and quite a few stayed open throughout the disaster. I was planning to stay in one of the resorts, as I felt that tourism is important for the Maldivian economy and that, by staying in resorts, we could contribute to rebuilding the economy.

It seems that visiting the Maldives as a tourist is probably the best way to contribute to the recovery effort. So I was trying to find out which was the best resort to visit, and I found that twenty out of a total of eighty-seven were completely destroyed. According to the tour company, one of the best resorts to go to was the Full Moon beach resort, which we booked in February 2005. This resort was also affected as a result of the tsunami; however, reconstruction and refurbishment of the hotel rooms and resort was ongoing, and it would be fully ready when we arrived at the islands in May 2005. At the beginning of May, I was informed that seventy-two of the eighty-seven resorts, including the Full Moon beach resort, were open for business, but twelve resorts remained closed. When my wife and I were on the Full Moon resort island at the end of May, we still found that the over-water bungalows were closed and some of them were completely destroyed. Rebuilding and repair works were ongoing.

Before I left for the Maldives, I went to a travel clinic to get my vaccinations and immunisation status updated, and I found that I had to take typhoid and hepatitis A vaccinations, which I duly did. However, I was surprised when the travel clinic nurse told me that I did not need to take anti-malaria medication. Later on, I found out from a Maldivian that, five years ago, the World Health Organisation (WHO) declared the Maldives to be a malaria-free zone.

As we were going to a post-tsunami area, I was prepared with all the appropriate emergency kits. It was important that I was particular about what I ate and drank. Usually, the chances of contamination are still present. Around sixty-nine inhabited islands lost their entire reserves of water. In some places, beaches were polluted. We also took lots of bottled drinking water. I was planning to take some antibiotics with my emergency kits, but I did not and this caused some inconvenience to me in the later part of our journey. When we arrived at the Full Moon resort, we were surprised to see many facilities and I found the island recovering quickly, although construction work was ongoing and the numbers of tourists were fewer. Plenty of food was available, as were water and other drinks. Essential services, food and water supplies had been quickly restored in most of the resort islands. In our resort, I was told that food, water and drinks were normally imported from outside, mainly Australia and South Africa, and I should not worry about food handling and its preparation.

The Maldives is famous for snorkelling and diving. There are beautiful sites and locations which are easily accessible, and we visited one of the uninhabited islands for snorkelling. While snorkelling, I saw dead and living coral and various marine creatures, including baitfish, but I was particularly interested to see the bleaching effect on coral polyps.

This bleaching is a kind of disease that affects the healthy reefs if the ocean temperatures rise too high. In 1998, there was a bleaching epidemic when coral mortality reached 70 per cent in some places, affecting the local tuna fishing economy. The recent tsunami is the second blow to their local economy, from which it is slowly recovering, and I saw this when I visited the local markets, including the lively fish market at Male. Although global warming will raise sea levels, coral reefs will survive as long as the oceans' temperatures do not continue to rise. One of the reasons for the less intensive effect of the tsunami on the Maldives is that the coral reefs acted as a barrier – another reason why the survival of the coral reefs is so important for them.

Some of the corals and marine creatures are dangerous to human beings. I first noticed this when I visited Jordan and the Red Sea (the Gulf of Aquaba) in February 2010.The Red Sea is part of the Great Rift valley and it was formed by Arabia splitting from Africa due to the movement of the Red Sea Rift. When I swam in the sea, I found that the water was warm and comfortable; the salinity was high but there was very good underwater visibility. The water was so clear that I was able to see some of the Red Sea marine creatures easily with my naked eye. The recent report on the Red sea's surface water temperature is 28°C in winter, increasing to 34°C in the summer. The salinity is high due to high levels of evaporation and the average is 40(ppt). The coral is healthy and there is very little sign of coral bleaching which might be due to coral having adapted to the heat.

While swimming on my own I was twice struck by fish I was not able to identify. The worst incident occurred when I was stung by an unknown marine creature on my right heel as I stood on it.

In the Red Sea, there are more than one thousand invertebrates and two hundred soft to hard corals. It is not easy to identify which one is which. However, it was a painful sting and I had to abandon my swimming and return to the shore where I was not able to stand on my right foot. On the beach my wife was waiting. She looked at my heel and found four puncture marks. At the same time nearby two lifeguards caught a Pufferfish. My wife shouted,' This is the fish which might have bitten you'. She is interested in marine biology as she studied Zoology at University. The lifeguards tried to kill the Pufferfish but my wife asked them not to do it. They honoured her request and let it go. As soon as it was back in the seawater it started swimming very fast and reduced its size until it was a four inch diameter ball shape; it submerged itself in the water and swam to the bottom of the sea. My wife tried to explain: 'The Pufferfish is a kind of poisonous fish that hides itself in between the rocks and corals. The masked puffer is usually small in size, but it attacks its pay, it releases a toxin and becomes bigger in size until gradually returning to its normal shape.'

Later on I did some research on the internet and found out that the Pufferfish is well-known for its unique defensive methods and is the second most poisonous vertebrate in the world. It has the remarkable quality of being able to expand its external body quickly when confronted with danger, revealing long poisonous spikes. It can grow up to sixty centimetres in length, though this depends on the type of species. Pufferfish mainly feed on the algae that grow on the rocks and coral and also the invertebrates that inhabit that area. It accumulates a neurotoxin called tetrodotoxin in its skin and internal organs. The skin and certain internal organs are highly toxic to humans though some of their flesh is considered to be a delicacy in countries such as Japan and Korea. Fatalities are not uncommon if it is not properly cooked and eaten. One of the doctors whom we met there described to me how she had come across a case of someone killed by eating Pufferfish. So it is most unlikely that I was bitten by a Pufferfish.

Some said it could be sea anemone. Another possibility is that I might have stood on fire coral. Fire coral is not real coral although it looks like the normal type of coral. It is cnidaria phylum which grows bush-like in sallow water on rocks and corals resembling seaweed. Fire coral has tiny tentacles which penetrate bare skin with venom that results in stinging pain. Whatever the creature, soon after the injury occurred, I applied antiseptic cream and took a painkiller, anti-histamine tablets and also a steroid but pain was persisting.

When I returned home my general practitioner (GP) decided to explore it through minor surgery and this helped me to recover. As a result, I learnt that Red Sea species are hazardous to humans and one should not swim or snorkel or drive without a proper guide who is familiar with the marine creatures, their danger and location.

Male, the capital of the Maldives, is about two kilometres long and one kilometre wide, and is the most populated island of the country. The rates of growth and population density are causing problems. Land reclamation was ongoing but has reached its limits, and a project for a new island is underway. Land reclamation has doubled the size of the original island and my local guide took me to the site of the recent reclamation areas, which have provided space for sports fields, outdoor recreation areas and artificial beaches, where I saw local boys and girls playing and swimming. To protect the island, my local guide showed me tetrapods (interlocking concrete blocks), which were built with Japanese help. Sea walls and breakwaters protect Male, and many people believe that, therefore, the effects of the tsunami were less severe on Male. It seems that the sea walls and coral reef barriers protected the island from the worst effects of the tsunami, although seawater still flooded half of the island.

In Male, I also visited the local market, where I tasted the local sweets, but I had to cast doubt on their hygiene, which prevented me from buying any. The day I left the Maldives, I felt sick at the airport. I had diarrhoea and vomiting. When I arrived home, I found that I had serious food poisoning, although my wife escaped it. To get rid of this, I took a course of antibiotics, but I was not sure where I had got this infection. Was it in the resort where I stayed or was it in Male? Of course, an infection like this is not uncommon after such a natural disaster. However, one has to be vigilant in order to prevent this type of occurrence. No doubt the Maldives is recovering from the disaster and hoping for greater tourist flow so that economically the country is able to survive by tourism as well as the fishing industry – bread and butter for Maldivians.

The tsunami hit southern India along the Coromandal coast, killing thousands of people. The most affected areas were Nagappatinam and Cuddalore in Tamilnadu. The government estimated that 30,000 homes in Pondicherry State and 100,000 families' homes in Tamilnadu State were destroyed.

The mighty tsunami also unearthed an ancient undiscovered temple at Mahabalipuram in Tamilnadu, and this was excavated after the tsunami hit the beach near the existing Shore Temple in Mahabalipuram. So I was eager not to miss the chance of visiting Tamilnadu when I went to India in August 2005.

In Kolkata, I contacted Rajib. Rajib Beed was a former student of St Xavier's Collegiate School, as well as its college, in Kolkata. Its alumni association was involved in tsunami-related resettlement and rehabilitation projects.

They gave me some contact numbers so that, if necessary, I could contact them when I would be in Tamilnadu. I flew from Kolkata to Chennai airport and it was a comfortable journey. I was pleased to see that aviation has developed considerably in India. Many aircrafts, both public and private, now connect the various cities of India. Nowadays, it is easy to buy tickets and seats are readily available, in contrast to my earlier experiences.

I arrived in time at Chennai airport, where my taxi was waiting. My next journey was Chennai to Pondicherry, where I stayed for a couple of nights.

Initially, I was scared to travel at night but I found the taxi driver who was sent for me to be reliable, and I reached Pondicherry safely at midnight without any difficulty.

The three-and-a-half-hour journey was mostly on the highway and I was pleased to see the development of a highway road. The roads were dual carriageways and, like Europe, there was a road toll system. India will develop a tremendous infrastructure if such highways (with the appropriate maintenance) connect all major cities of India. Since 1999, India has been working on the Golden Quadrilateral road project, which links the country's four major cities (Delhi, Mumbai, Chennai and Kolkata) with 5,850 kilometres (3,630 miles) of highway.

In Pondicherry, I stayed at the Sea Side guesthouse, comfortable accommodation facing the sea. In the morning, I was very pleased to see Bokul and his wife, and for the next two days Bokul Ghosh was my companion. They live in Pondicherry. I first met Bokul in my teenage years at Chandranath's home in south Calcutta. Chandranath Dey was one of my closest friends. He was a keen sportsman, especially at tennis, was good at socialising and always helpful to others. He came from one of the most renowned business families in Calcutta. His premature and sudden death in 1998 shocked me very much. Since then, whenever I go to Kolkata, I miss Chandranath, although his wife Sabrani and son Raja are still close to us. Raja now runs his own business, which manufactures motor batteries, in the Himalayan kingdom of Bhutan.

Without their help, my trip to the tsunami- affected area of southern India would not have been successful.

Pondicherry was the capital of French India and the French influence is still present. During the tsunami, the town of Pondicherry was virtually untouched, although many houses in nearby villages were destroyed. Most of the villages were fishing villages. The huts were rebuilt and I was told that each fisherman got a new boat, built locally using donated money from various countries of the world. I visited one such village where twenty people had died. It was nearly midday and the day was very hot, although there was a gentle breeze coming from the sea. Not many people were on the beach, except for some villagers. The fishermen, after the morning catch, had returned and their empty boats were lying on the beach. I could see the donor countries' emblems on the boat panels. The fishermen were resting and knitting their nets and women were counting money, chatting or sorting out their fish under the newly built shelters, with roofs made of casuarinas and fibreglass.

Fish were drying on the disused boats. Nearby in the village, the womenfolk were preparing fish to be cooked. A naked child was moving around; however, I did not see many children on the beach. Whoever I met in that village looked happy and smiling. It was hard to believe that the tsunami had had an impact on this village. This had to be one of the villages that were settled quickly.

In the evening, I was strolling around with Bokul on Pondicherry's main seafront and we stopped in front of the Pondicherry tennis court. Bokul told me, 'The day the tsunami hit Pondicherry, no other area of the city of Pondicherry was affected except this tennis court. We all came to the tennis court as sea waves hit it and left it full of marine creatures as soon as they receded.'

The next day, I visited the newly discovered ruins of the temple in Mahabalipuram, after travelling along the east coast road from Pondicherry. On the way, I passed through the fishing villages, sandy beaches, barren fields and shoreline and roadside huts that were once the temporary shelters for tsunami victims. I saw the backwater lagoons and pools along the shoreline, where sea salts are manufactured from the natural ocean water by the process of evaporation. The tsunami also affected the sea salt- making industry; the worst affected area was forty-five kilometres south of Nagappatinam, where the tsunami washed away thousands of tonnes of stock salts that were then contaminated with debris and black silt.

Along the coast, I saw beautiful sandy beaches but very little vegetation, except for the occasional glimpse of coconut, palm and casuarina trees; one can see a similar scene throughout the Coromandel coastline. I was told that, a thousand years ago, there was vegetation, but deforestation and land erosion were responsible for the present situation. After the tsunami, there was a debate over whether sea walls along the Coromandel Coast or forestation with appropriate plant species would best protect the coastline from natural disasters like this. I spoke to Prabir Banerjee, president of Shuddham, an organisation based at Pondicherry whose aim is to beautify India by recycling domestic waste and reviving the ecosystem by reforestation. Their plan is to plant trees along the Coromandel Coast, and he said that an Australian, resident in Auroville, had identified certain plant species that survive tsunamis.Auroville has also developed village-based programmes to teach about the Tropical Dry Evergreen Forest (TDEF) ecosystem. I also visited Auroville in Pondicherry; near the Tamilnadu border.Auroville is a place, spread over twenty kilometres, where men, women and children from various parts of the world live in peace and progressive harmony with each other, irrespective of creed, politics and nationality. It is an experiment in international living, which was founded in February 1968 by the Mother, a French woman, one of the devotees of Sri Aurobindo, founder of the Aurobindo Ashram in Pondicherry.

Mahabalipuram, about fifty kilometres from Chennai, is world-famous for its Shore Temple. There are also other temples, like rock-cut cave temples, monolithic rathas and structural temples. The Pallava kings built most of the monuments between the fifth and eighth centuries AD.

Mahabalipuram was their seaport and they ruled from the capital, Kanchipuram, in southern India. European travellers who came to Mahabalipuram in the fifteenth and sixteenth centuries described the 'Seven Pagodas' on the shore at Mahabalipuram, but their existence is debatable because of a lack of archaeological evidence. However, after the tsunami, there has been fresh light on the concept. When the waves first receded about 500 metres into the sea before the tsunami struck the Mahabalipuram coast on 26 December 2004, tourists and local people saw a row of rocks and some architectural remains of a temple on the seabed. When the waves subsided, these were submerged in the sea again. When the tsunami waves receded, they washed away a vast quantity of sand from the beach into the sea; people saw the rocks on a square area south of the Shore Temple. The remains of the temple were unearthed fully, following further excavations by the Archaeological Survey of India (ASI). I visited the site and found the twenty -metre- wide and twenty-five-metre-long collapsed temple, and saw the terracotta ring well, which was found among the ruins.

According to archaeologists, the discovery of terracotta well in the temple complex supports the theory that it belongs to the earlier period and not the Pallava period. The ruined temple was not fully open to the public and excavation was still ongoing. To the east, out to sea, offshore underwater exploration took place in February 2005 on the seabed and divers saw walls between the rocks. This supports its existence, but further investigation is needed into the submerged structure of the temple.

The local guide took me to the Tiger Cave, four kilometres north of Mahabalipuram near the sea shore, where some archaeological excavation was happening, not open to the public. I met some of the archaeologists from the ASI who were excavating, and they allowed me to see the site. One of the archaeologists, Mr G Thirumoorthy, showed me a terracotta lamb that had just been excavated. Other items that were found included terracotta roofing tiles and conical pots. He also pointed out to me brick-made stairs on the excavated site, unusual for the temples of Mahabalipuram. This suggests the temple was most probably built in the pre-Pallava era, that is, the third to fourth centuries AD. He allowed me to take some photographs.

Since the tsunami, the whole of Mahabalipuram is now an exciting site, especially for the historian, archaeologist and studier of temples. Further investigation will no doubt answer many questions and reveal further archaeological discoveries.

After travelling twenty miles on the east coast road, I saw two types of temporary shelter for the tsunami survivors. On the left side of the road, there were square- or box-shaped blue huts, made of corrugated tin or fibreglass; on the right side of the road were huts, made up of casuarinas.

People were still living in the temporary shelters made of casuarinas' poles and roofs, but none lived in the box-shaped huts. I was told that people had abandoned the box-type accommodation because they were too hot to live in, and they found casuarinas' huts more comfortable, especially in summer. I could not see any tents now that had been put up after the tsunami. There were settlement colonies built away from the coast with proper toilets, bathrooms and water facilities, but half of them were not occupied. This was because most of the allottees had returned to tsunami-damaged coastal homes, as they were unable to live so far away from the sea.

My tour was hectic. I stayed two days and as such I was not able to cover all of the tsunami-affected areas, and while, no doubt, most of the places had recovered, the memories were still alive.

Earthquakes trigger landslides; the classic example was the 8 October 2005 earthquake in South Asia. It was a memorable day, as it was a weekend and we were going to a social function in London. Instead, we had to proceed to Heathrow airport, as Rahul was coming from Madrid, where he and Rumella had been for a friend's wedding. Rahul had to go to Pakistan to cover the earthquake disaster. He had to leave for Islamabad on the evening flight, and had very little time. Within two or three hours, he had to collect tickets, a visa and money, go to the BBC office and then home, to pack up the essential items to cover a disaster zone as a journalist. Without our help, he would not have been able to catch the flight in time, as nobody was in his London home and Rumella was still in Madrid. It was lucky that he already had a Pakistani visa and so did not have to go to the Pakistani High Commissioner's office this time. I advised him regarding some tinned foods, water and medicine, and my wife gave him a sleeping bag in case he had to live rough.

He was able to catch the late evening flight to Islamabad. Rahul was in the Pakistani capital on the next day and reached the most devastated area, the town of Bagh, nearly fifty kilometres from Islamabad, on

11 October 2005. It took Rahul and his team seven hours to travel a distance that normally takes three hours. Rahul Tandon was one of the first journalists to reach the town, three days after the incident. He found that 80 per cent of the village was destroyed and no children had been found alive. A girls' college had collapsed, with 300 bodies unrecovered and no equipment available to remove the bodies. He reported on this for the BBC World Today programme.

Rahul also visited remote places, where relief efforts had not reached, and he saw a child survivor who later died from the cold weather. There were some outlying areas, higher up on the mountain, where the weather was already cold and getting colder as winter was approaching. People were moving around on the street as there was nowhere to stay, and some were wearing facemasks because of the large numbers of dead bodies. They were desperate for some sort of aid to arrive.

Rahul stayed for twelve days in the disaster zone and, from Islamabad, he went to places like Bagh, Muzaffarabad and Balakot, and also covered some of the surrounding villages. When he returned home, I asked him, 'Were you involved in relief work apart from your journalistic work?'

He answered, 'Not a lot, but I shared my food, medicine and the water that I had carried from England, and I also donated my sleeping bag to an earthquake victim.' He said, 'We met a fifteen -year- old boy on the road near Bagh who had been walking for seven hours, carrying medicine on his back to his village on the hilltop for the earthquake victims. He was tired and exhausted, so we gave him a lift.'

Rahul also said, 'Prior to the arrival of aid, whatever support we were able to give by our small means, we gave.'

I said, 'That is right, but it is more important that as a journalist you were able to focus the world's attention by reporting the facts from the earthquake zone, thereby promoting international aid and the relief operation.'

The epicentre was about three miles from Balakot in the Nellam Valley of Kashmir, and the magnitude was 7.6 on the Richter Scale. Its impact was felt in Pakistan, northern India and Afghanistan, although the main affected areas were Kashmir, both Pakistan- and Indian-administered, and the north-west frontier province of Pakistan. The earthquake killed more than 73,000 people, injured 60,000 and left 3.3 million homeless. Landslides buried many villages and roads, and towns and infrastructure were destroyed. Moreover, the difficult mountainous terrain and worsening weather conditions made it the hardest disaster to respond to in recent history. Problems with access delayed the arrival of rescue teams, resulting in the delaying of treatment and basic help to the victims. According to the WHO, twenty-six hospitals in Pakistan were destroyed. Death, disease and disability rose due to delayed treatment, resulting from damage to roads, hospitals, water and sanitation, and poor transport facilities. Lack of tents forced homeless people to sleep outside in cold and harsh weather conditions.

The doctors found that many victims arrived too late, either dead or suffering from gangrenous wounds, resulting in amputation. The number of limb amputations increased drastically, especially in children, requiring long-term costly rehabilitation. The international rescue effort was severely hampered by treacherous mountain terrain and huge landslides. Lack of heavy machinery and drills forced people to rely on hammers and shovels, resulting in a slow and painful rescue operation. There were many British people of Pakistani or Kashmiri origin who lost their relatives in the earthquake. Many flew to Pakistan to help relatives in the disaster zones. Britain sent a specialist rescue team, money and aid. Other major countries of the world offered search and rescue teams, money, aid, tents, helicopters, doctors and nurses. Even India offered help. The devastation of the earthquake affected the Line of Control, killing army personnel and civilians, and damaging property. Aid effort crossing the Line of Control, which divides Kashmir into Indian-and Pakistani-controlled areas, was blocked. However, the earthquake in Kashmir improved the relationship between India and Pakistan by opening up certain border posts so that Kashmiris could cross the border, see relatives and help the victims and survivors.

Landslides can cause a phenomenon known as 'Quake Lake'. The quake lake is usually formed on rivers which are blocked by large landslides following an earthquake. The typical example is the Sichuan province earthquake which occurred on the 12th May 2008 in China. The epicentre was eighty kilometres north-west of Chengdu, the provincial capital of Sichuan province, at a depth of nineteen kilometres. The magnitude was 8.0 and there were many aftershocks. It was the deadliest earthquake in China since the 1976 Tangshan earthquake.

According to the Chinese Government the earthquake killed 69,181 people, 374,176 injured and 18,498 are in missing. More then ten million people were affected by the disaster. Millions of live stock and agriculture were destroyed. Schools, factories, many houses, flats and hospitals were reduced to rubble or damaged. Substandard cement and shoddy construction were blamed for the heavy casualties and number of fatalities. Those who survived were often left homeless and faced disaster- related health problems. According to one source, total damages exceed US $ 20 billion. The Zipingpu dam and Wolong panda reserve, which were not far from epicentre, were affected. Across the whole region I gather that a total of four hundred dams, both new and old were damaged. Nineteen British tourists who were trapped in the giant pandas reserve forest area escaped unhurt, although three giant pandas were reported missing. China tried to tackle this catastrophe themselves but it was too great a task so they called for outside assistance. Many countries like Japan, South Korea, Singapore, Russia, Taiwan, USA, and UK extended their help in rescue operations or relief work. Donations from overseas exceed US$860 million within a week of the earthquake. The World Health Organization and the Red cross were also involved. The Chinese authorities avoided a major outbreak of disease by dispatching medical staff, setting up field hospitals and supplying disease prevention materials in time.

Within fifteen days of the initial earthquake about thirty-four 'quake lakes' were formed due to earthquake debris blocking and damming the rivers. Eighty-two percent of these lakes were a potentials danger to local people because of risk of flooding and the dam busting. Rising water and heavy rain did not help either. The largest and most dangerous 'quake lake' was on the Tangjiashan Mountain. To avoid further catastrophe, 250,000 people were evacuated downstream to the valley below and the bottle-necked water from the quake lakes were discharged by blasting. This was achieved by using anti-tank rockets and dynamite by the army. As soon as this dangerously unstable quake lake was blasted, a torrent of muddy water and debris along with dead bodies, cars, and household goods was swept downstream through the controlled flood.

Earthquakes can trigger volcanic eruptions. On 27 May 2006, nearly 6,000 people died and 340,000 people were made homeless by an earthquake measuring 6.2 on the Richter Scale, near the city of Yogyakarta, Indonesia. The earthquake's shockwaves disturbed the magma bubbling inside the nearby 9,800-foot volcanic mountain, Mount Merapi. Frequent tremors, lava flows and clouds of gas indicated that a significant eruption was about to occur. The last major volcanic eruption of Mount Merapi had been in 1994.

Earthquakes or volcanic eruptions occur on the earth's fault lines due to the movements of the boundaries of the earth's tectonic plates. The tectonic plates, are:

1. North American
2. Pacific
3. Caribbean
4. Cocos
5. Nazca
6. South American
7. Scotia
8. African
9. Arabian
10. Eurasian
11. Philippines
12. Indian or Indo-Australian
13. Antarctic
14. Iranian
15. Cardine
16. Fiji

There are ongoing disagreements among scientists about the exact number of plates, and about whether particular named plates constitute one plate or a series of smaller ones. For example, there is a debate over whether the Indian and Australian plates are one or two.

There are three types of plate boundary:

1. Transform boundary: the plates slide or grind along the fault lines.
2. Divergent boundary: the two plates pull away from each other.
3. Convergent boundary: the plates push towards each other, resulting in one plate moving under the other, or the plates colliding with each other.

The focal point is the epicentre, where the most severe impact of the earthquake occurs; shockwaves get weaker as they spread out from that point. The effects of the earthquake might be primary and secondary. The primary effects are on land, trees, structures, animals and people. Devastating effects are noticed more in the city than the countryside. The secondary effects of earthquakes are fire, tsunamis, landslides, avalanches, and change of land scenarios.

The effects of earthquakes on human health and their cost on population are often high. Serious injury or death is caused mainly by the collapse of buildings and other man-made structures, such as roads and bridges.

The death toll is usually highest in cities where there are tall buildings, badly built flats and houses, and areas of dense population. In recent earthquakes in Sichuan, China (2008),in Kashmir (2005), Gujarat in India (2001) and Izmit in Turkey (1999), badly structured houses and flats were blamed for the heavy casualties.

There is no correlation between the magnitude of the earthquake and the death toll. The global history of earthquakes shows that there might be a high-magnitude earthquake where the death toll is nil. On the other hand, there may be heavy death tolls where the magnitude is low. The number of people killed does not always correlate with the magnitude and severity of the quake.

Two classic examples in this century are the Chilean earthquake that occurred on the 27 February 2010 and the Haitian earthquake on 12

January 2010.The Chilean earthquake was stronger but the death toll of Haiti's earthquake was significantly higher.

The magnitude of Haiti's earthquake was 7.2 on the Richter scale followed by a number of aftershocks which killed 233,000 people, injured 300,000 and left one million people homeless. Three million people in total were affected. The epicentre was fourteen miles west of Port-au- Prince, the capital of Haiti, which is densely populated. It has been estimated that 250,000 residential and 30,000 commercial buildings were severely damaged or collapsed. Many famous buildings including the Presidential buildings and the headquarters of the United Nation Stabilisation Mission in Haiti were also damaged or destroyed. Aid workers, certain embassy staff, tourists and eminent personalities were amongst the dead. Haiti is one of the poorest countries in the world and natural disasters such as earthquakes, hurricane-related storms and floods are common, which have a great impact on the nation's economy.

Chile's earthquake measured 8.8 on the Richter scale and was followed by aftershocks and tsunamis but the fatalities numbered around 796. Chile is richer and has strict building regulation codes and a better emergency response infrastructure than Haiti, hence it is no wonder that the earthquake in Haiti had more devastating effects than the earthquake in Chile.

Chile lies in one of the most active tectonic zones of the world and earthquakes are common. The biggest earthquake of the world which has been ever recorded occurred in Chile in the year 1960;it measured 9.5 on the Richter scale and also generated tsunamis which affected southern Chile, Hawaii, Japan, the Philippines, New Zealand, Australia, Aleutian islands and Alaska. The total numbers of fatalities of the 1960 earthquake are not fully known but on the basis of several studies it has been estimated that most probably between 2,231 and 6,000 people were killed.

The Chile's recent earthquake (27 February 2010) struck along the boundary between the Nazca and South American tectonic plate. The location is twenty -one miles(thirty-four kilometres)deep in the sea at the convergent boundary where the subduction process is occurring at a rate of eighty millimetres (three inches)a year. According to NASA's scientists, the Earth's axis may have shifted and also caused the day to shorten as a result of this Chilean earthquake.

It is interesting to note that when I visited Assam, India, in 1967, I found that most of the houses in Assam were built of wood. I was told that Assam is prone to recurrent earthquakes and so the houses were constructed in such a way that there would be fewer casualties.

Following an earthquake, the first necessary step is a rescue operation, followed by food, shelter, treatment and prevention of epidemic and disease. There may be issues of law and order including looting and violence, which need to be tackled immediately. Displacement, disease, famine and epidemic are the after-effects. Psychological effects take the form of shock, fear and panic attacks; bereavement and post-traumatic stress disorders are not uncommon among the survivors. Rehabilitation is important.

Future efforts in preventing earthquakes throughout the world depend upon prediction, damage limitation and handling. No doubt the world's scientists, architects and engineers have great challenges ahead. Already, some scientists have claimed that they have discovered a method of predicting and even preventing earthquakes from occurring. The idea is that measuring stations are placed along a fault line equipped with special microphones named 'geophones', which detect the low frequency noise generated by the earth's tectonic plate movements. Scientists will be able to determine the area of a potential quake and then, by drilling deep below the surface, extract a sample of rock to evaluate its strength, size and shape. This will help to find out how much stress the rock can take before it gives way. Lasers can then be used to read tectonic movements, to work out how much force the plates exert. A deep ultrasound survey will be able to locate the specific rock where the pressure has accumulated. Using this information, scientists can predict when the obstructing rock will give way, causing an earthquake. To prevent this, a hole can be drilled into the rock (up to ten kilometres deep) and explosives placed within it, followed by a series of controlled blasts that weaken the rock, gradually relieving the built-up energy and thus preventing a major earthquake. This concept is still in the realm of ideas, but hopefully one day scientists will be able to materialise it and, if engineers and architects are able to prevent the devastating effects of the earthquake by constructing more earthquake-resistant structures, houses, roads and bridges, the terrible effects of earthquakes can be neutralised.

Andaman's landscape following tsunami and earthquake, December 2004

Chapter Two
Volcanoes

Volcanic plume from an eruption in Hawaii.

Volcanic eruptions fascinate me. The spectacular views of their activities can be captured by photograph and are heaven for artists and photographers. The classic example was the eruption of Grímsvötn, Iceland, in 1996, impressive aerial photography of which was televised throughout the world.

Some say that volcanic eruptions are the anger of Mother Earth. The ancient Romans believed that Vulcan, the god of fire, lived deep inside a mountain in the Aeolian Islands. According to Hawaiian mythology, volcanic activity is caused by the fire goddess, Pele. In Central America, one tribe used to offer human sacrifices to the crater to stop volcanic eruptions. Since ancient times, various attempts have been made by natural and supernatural means to stop volcanic eruption. Whatever the mythology is, scientifically volcanic eruption is certainly dramatic and unpredictable. The nature of each eruption is dependent upon the composition of the earth's interior.

This is based upon the geological, geophysical and geochemical make-up of the earth's surface; that is, the crust, the inner core and the outer core. Volcanic eruptions vary in form from hot water to molten lava and from the finest ash to rock debris. There are also secondary effects, contributing to the constantly changing pattern of our earth.

Scientists have categorised volcanoes on the basis of eruption styles:

a) Mild: The eruption is short-lived, of small extent, fragile and non-threatening. It is usually seen in the form of hot springs and pools, without lava formation.

b) Moderate: Cinders and cinder cones, fissures, bursts, ash, huge shields and lava flows.

c) Vigorous: Ash, cinders, blocks, noise, ash cones, sediment, blasts and gas eruptions.

d) Violent: Ash, floods, domes, pumice, blasts, landslides, avalanches, hot clouds of air, hot gasses and aerosol formation.

e) Supervolcanic eruption: At least one thousand cubic metres of magma is expelled.

Volcanic mudflows, tsunamis and a change in the earth's landscape are also major side effects of volcanic eruptions.

Presently, some of the volcanoes of the world are in the eruptive stage and some of them are in the dormant stage.

The geological timescale of volcanic eruption is certainly different from human timescales. However, the historical record of major volcanic eruptions goes back to 1500 BC in Santorini in Greece and Mount Etna in Sicily, Italy.

Santorini is believed to be the lost continent of Atlantis in ancient Greek mythology. In 1545 BC, a catastrophic, titanic eruption took place that destroyed the local settlements of the island and created a tsunami that wiped out the Minoan civilisation on Crete, more than seventy miles away. Santorini is still active and it is one of the few active volcanoes in southern Europe, other than in Italy.

Italy is a country where all types of volcanoes can be found. People in this part of the world have to live with ongoing and active geological processes.

Etna has the longest history of documented eruptions of any volcano since 1500 BC. It is Europe's largest volcano and one of the most active volcanoes in the world. The intermittent to persistent activity occurs frequently in the summit area and major explosive activity occurs occasionally, although the main feature of Mount Etna's activity is lava emission. In the new millennium, flank eruption has occurred four times, in 2001, in 2002–2003, in 2004 and in 2007. In October 2004, during my visit to Sicily, I found the eruption still continuing, having begun in mid- September 2004. I saw the snow-covered summit of Mount Etna and gas plumes rising from the crater.

The average sulphur dioxide flow in the month of October 2004 was 3,400 tonnes per day as the lava continued to flow; tourists, scientists, volcanologists, artists and photographers flocked in for a glimpse of the eruption. In 2002–2003, the eruptive activities of Mount Etna produced lava fountains and explosive activity (bomb ejection), damaging the access road and destroying the former tourist station. Moreover, hundreds of Italians fled from their homes, as streams of lava rushed down the north-eastern and southern flanks of the mountain. Mount Etna has erupted many times over the past 2,000 years; the eruption of AD 1669 destroyed the city of Catania, leaving the citizens starving. Boiling lava covered most of Catania, leaving the city with ruined streets and uninhabitable buildings, on which the modern attractive city was built. On the way from the airport to Catania, one can still see the dark lava stone structures that witnessed the devastating past effects of Mount Etna.

In January 2006, I visited Sicily again. This time, I was with my whole family for a winter break. The whole family means me and my wife, my two daughters, grandson and son-in-law Rahul. Ashish Sharma, my eldest daughter's colleague at the BBC, also joined us at Taormina. My last visit to Mount Etna had been in October 2004. That time, I saw Etna from Catania, which is in the south, and this time we saw it from Taormina, which is in the north and was where we stayed for two days. Last time, Etna was erupting and this time it was in a dormant state. However, we were not able to go to the top because of snowfall, but we were able to do the same tour of the surrounding area by car. Narrow roads, poor traffic sense and irregular parking made driving in Sicily sometimes difficult and miserable. We saw the recent basalt lava, which was once fluid lava and looked like rivers of fire running through the mountain towards the valley with a temperature of more than 1,000°C. Our visit was in winter and some of the basalt lava we saw was covered with snow. From a distance, Mount Etna often looked white, as for seven months of the year it is usually covered with ice and snow.

When the snow melts, it feeds the River Alcantara, one of the rivers of Sicily, which begins at Mount Etna; we visited the gorges through which the Alcantara flows; its river bed is composed of basalt lava. The river and ash from Etna's eruptions makes the Randazzo and Taormina valley a fertile land, full of orange groves, hazel groves, vineyards and chestnuts. Power plants, upon which Sicily depends, are also fed by the Alcantara River. We visited a place called the Terremoti, where roads are built with volcanic rocks, and in the centre we saw a church that had been damaged by earthquake in July 2001, and was still awaiting repair. Earthquakes have been recorded near Etna in the same area that was affected by the volcanic eruption.

Mount Etna erupted again in September 2007. I was not able to go, but my younger daughter Tanya and her husband, Andrew Hudson, were in Sicily on holiday. They went up Mount Etna on 3 September and, at 2,200 metres, saw eruptions in the form of smoke and bomb ejections on areas within one hundred metres. From there, they could not see any lava flow activity. They were not allowed to go close to the eruption site, but were able to take photographs of this activity, which they shared with me.

In October 2004, my wife and I visited the volcanic islands in the Mediterranean Sea: the Aeolian Islands, Isole Eolie in Italian. They consist of seven large and a number of small to very small islands. The Aeolian Islands emerge from the sea in various sizes and heights. They belong to a unique volcanic group of tectonic environments that are related to subduction, although they all are different in their character and colours.

The seven major islands are Alicudi, Filicudi, Salina, Lipari, Vulcano, Panarea and Stromboli, and all are located off the northeastern coast of Sicily.

We stayed for eight days in a hotel at Lipari and visited each island on different days, swam in the sea, climbed up to the volcanic craters and enjoyed the Aeolian cuisine. It was the most memorable time I have ever had.

Panarea is the smallest island, and Lipari is the largest and most populated. Alicudi, Filicudi and Salina are totally dormant, while Lipari and Panarea still have some thermal sources and fumaroles. Stromboli and Vulcano are still active.

Many believe that the Aeolian volcanism started not more than one million years ago. The islands of Alicudi, Filicudi, Panarea, and parts of Salina and Lipari, were created in the first phase of the activity and, during the second phase, Salina and Lipari were completed, and Vulcano and Stromboli were created.

The island of Vulcano is believed to be around 100,000 years old and the little island of Vulcanello, created in 183 BC, is now attached to Vulcano. Stromboli has a history of continuous or frequent eruptions for at least 3,000 years. While swimming in certain areas of the sea there, one realises that there is still underwater volcanic activity, and there are incidences of adverse effects. Fishermen have reported feeling dopey while fishing and young lovers felt dizzy while boating around the Battaro, near Panarea, leading to the discovery of underwater gas emissions. This unique phenomenon of underwater fumaroles we also noticed when visiting the area around the Battaro. On another occasion, we abandoned our swimming around a cave near Filicudi, because of underwater fumaroles and marine creatures, including jellyfish.

Scientists have already analysed the gas emissions around Panarea. Some of the flows consist of more than 200,000 litres per day of gas composed of carbon dioxide, sulphur, methane and hydrogen, and some are surrounded by enormous deposits of white sulphur.

Stromboli, the ancient Strongyle, meaning 'round', is one of the most active volcanoes on Earth and has been continuously erupting for thousands of years.

The local people say that Efesto, the god of volcanoes, never leaves this island. We visited Stromboli on 27 October 2004.There are two landing places on this volcanic island. One is on the south-western side called Pertuso (from the Latin pertugio) in the village of Ginostra; the other is to the east, called Scari (meaning 'port'), in the village of San Vincenzo. We landed in Scari on the afternoon of the 27 October after a tour around the island by boat. While sailing around the island, we saw the lighthouse, deep blue sea, black sandy beaches and a wide black slide-shaped hollow where the debris and lava from the volcanic explosions flow between the two rocky walls. On landing at Scari, we climbed up towards the church of St Vincent along a narrow road. It was built as a small shrine in 1615 and became a church in 1725, which has since been restored and enlarged. From this point, one can see the whole village and, at the church, a track leads to the top of the volcano. On the narrow road, I saw a road sign indicating that children should wear gas masks. It reminded me that, although violent eruptions are rare at Stromboli, large and unexpected eruptions do occur. Over the past ten years, several people have been injured or killed due to accidents and explosions.

It was getting dark and we had to climb down, sampling a pizza in a local restaurant before we got into the boat for a view of Stromboli's eruption at night from the sea. After sailing out on a boat for twenty minutes from the north-east corner of the island, we had our first glimpse of the Stromboli eruption, consisting of small gas explosions that hurled incandescent blobs of lava above the crater rim.

This spattering, lasting for a few seconds, occurred at intervals of between five and twenty minutes. We were on the upper deck of a small boat called Principessa and I was waiting patiently for a splendid large eruption of Stromboli on that dark windy night, although the full moon was the next day. The boat moved up and down, passengers were swaying from left to right, and I felt that the boat was being tossed like a yo-yo on the stormy waves of the Mediterranean Sea. After observing five or six small erupting phenomena of Stromboli, we had to abandon our position on the top deck as the captain of the boat, Angelo, announced that a big storm was coming and we should go back to the shore as fast as possible, and that everybody must take shelter in the enclosed deck below.

The boat was caught in the storm and later on the rain started; one third of the passengers, including myself, was sick or felt sick. The boat ran as fast as possible towards the shore but, even so, it took more than an hour to reach Lipari. The captain and crew were very nice; they took care of the sick passengers by providing the appropriate plastic bags and tissue papers. We certainly realised how ferocious the storm was, and the incident reminded me of the myth of Aeolus, who controlled the winds from this island; this is because in olden days the sailors were able to tell the direction and strength of the winds from the smoke of Stromboli. Perhaps Captain Angelo also predicted the strength of the storm by judging the smoke from Stromboli.

The other most active volcano in the Aeolian Islands is the island of Vulcano. Since its creation, considerable volcanic activities have occurred on this island, which we saw as we approached by ferry on 23 October 2004, on the way to Lipari from Milazzo, and again by small boat on 29 October, when we landed on Vulcano to climb the volcanic crater.

Sailing around the island, one can easily observe its various natural phenomena: the caves opened up by the lava flows; the beaches divided by lava promontories that stick out into the sea; the coastline with steep, rugged slopes and the crack of the crater where the smoke rises; and the most remarkable one, the Monster Valley, created from strange rock formations.

The rock formation reminded me of our visit to Cappadocia in Turkey in 2002. About ten million years ago, the eruptions from the three volcanoes (Erciyes Dagi near Kayseri, Hasan Dagi near Aksaray and Melendez Dagi near Nigde) produced a thick layer of hot volcanic ash that covered the region. Over the years, it hardened into a soft, porous stone called 'tuff'. With wind, water and sand erosion, the tuff wore away and curved into various shapes. On the top of the tuff, there were boulders and hard stones with a rather phallic appearance, and this type of column or cone of tuff is sometimes called a 'fairy chimney'. The day we arrived in Cappadocia, there was snowfall. However, the snow soon stopped; the sun was shining and the landscape was beautiful. The cave dwelling, carved out of a single cone of tuff, the connecting tunnels and the decorated, rock-cut churches are typical of Cappadocia. It is hard to believe that tuff is nothing but hard volcanic ash.

Similarly, I saw an interesting rock formation in Moon Valley, when we visited La Paz in Bolivia in July 2007. La Paz is the world's highest capital city, at an altitude of 4,000 metres (13,100 feet). Moon Valley is located south of La Paz, ten kilometres from the city centre. It appeared to me to be volcanic rock eroded by rainfall. When I touched and examined it, unlike the formations in Cappadocia, I found that this was not so hard or tough. It was fragile and was made of soft sandstone. I collected a sample for my own collection. The area is a glacial valley, but does not at present have glaciers. The glacier melted away millions of years ago, and has left bare, soft sandstone. The rain and wind have eroded the area into tall spines and crevices. Our local guide informed us that Neil Armstrong had given this area the name 'Moon Valley'. The erosion has given it the appearance of the surface of the moon (moonscape), although I am not fully convinced, as it seems to be more like numerous sandcastles on a series of steep valleys.

The American astronaut Neil Armstrong was the first person to land on the moon, which took place on 20 July 1969. I still remember how excited I was to be watching the moon landing event, which was televised live throughout the day and night.

Our small boat arrived at Porto di Levante, the main port of Vulcano Island. As soon we got down from the boat, we could smell the presence of sulphur in the air and could see yellow coloured rock formations, which reminded me that at one time nobody except slaves and forced labourers used to live on this island, for the extraction of sulphur and alum.

Climbing the crater was a great experience. From Porto di Levante, we walked along the road and, after 150 metres; we turned to the path on the left that leads to the crater. At the outset, we were stopped by a man and had to pay a fee to go up to view the crater. My wife did not want to go up and so she stopped at a nearby hut, a small tourist resting place. I started to climb, although I was not sure whether I would able to reach the top; health-wise, I was not one hundred per cent fit. At first, the climb was easy, and then the path got more difficult and uneven, and the track wound back on itself. I passed through the dark, ash coloured pathway, black-coloured sands, reddish rock and a barren landscape. As I climbed up, the scenery changed; I could see a valley with vegetation and vineyards, and also caught sight of the sea and other islands. I could also hear sounds that might have been volcanic explosions indicating ongoing volcanic eruption. At that moment, I was alone and had mixed feelings, fear as well as excitement, because of what I would see at the top of the crater. I was slow to climb and this was because of my limited eyesight and recent recovery from a cartilage operation. However, I climbed the 124 metres on my own to see the crater, and most of the people who passed by me said 'Hello'. Overcoming the last obstacle, I reached the crater's edge and saw some of our group members. I was fascinated to see the mouth of the crater. I stood at the top of the crater and saw the fumaroles, rich sulphuric crystallizations. Smoke and the intense yellow of sulphur were coming out of the crater's depth. I took a photograph and walked along the crater towards the north edge, where a straight crack had recently formed. There were holes with smoke and vapours pouring out, whistling loudly and saturated with sulphur.

It was a very irritating and suffocating gas that I could not tolerate for long. I felt choked and found difficulty in breathing, tears starting to pour out of my eyes. I pulled out my handkerchief from my pocket to use as a mask, even though I was not comfortable. The smoke and vapour were sometimes so dense that I could not see the pathway near the crack line. I had to leave the area to prevent further exposure. However, I felt that I should have carried some instruments to detect or to measure the gas and vapour emissions. Whatever the outcome, I felt that I had seen what I wanted to see and that the climb to the top of the crater had been worthwhile. It was now high time to go down. I also found that none of our group members was on the top of the crater. They might have already started to climb down.

I started to descend and, after passing barren land, the track became stiff and I found that it was more difficult to climb down than to climb up, although in front of me some of the children and elderly people were climbing down easily. I suddenly realised that my disability was getting in my way. The uneven track was making me unsteady, vulnerable to falling, and my walking stick was not good enough to support me. I felt that I would be trapped there for ever and not able to climb down. However, slowly, and nearly by crawling, I somehow managed the immediate problem of tripping over, but there was still long way to go. I felt great relief when I heard a female voice in broken English saying, 'Do you need any help?' I gladly accepted her offer and she supported my right side while I slowly climbed down. We started talking. The process of walking down became much easier when one had such physical support, as well as chatting, which we did.

The young girl said, 'We are supporting each other, as I have a phobia of heights.' I was rather surprised to hear this, as she quite often went to Scotland to walk on the Scottish hills. She was from Germany and she told me her name but it is difficult to remember, although it did not sound like a German name. She explained to me that her first name was Norwegian and the family name was Swiss. She worked in Germany in a community mental health team as a social worker. She had arrived in Sicily two days before and the previous day had climbed Mount Etna and saw its eruption. She also asked me whether I was in a group or on my own. I said, 'No, I am in a group but, as you see, none from my group, not even my wife, is with me.' We eventually descended and reached the tourist resting place where my wife was waiting and then the young girl left me, saying goodbye. She disappeared as quickly as she appeared, but my memory of her is still strong.

Before we left Vulcano Island, I had a bath in the mud pool but I was not impressed by it, as I found more gravel than mud, although the thermal baths of this place are renowned for their curative powers.

In the Aeolian Islands, I had hoped to see coral islands or an underwater view of corals and marine life, which I could not. As far as I know, coral islands begin life as a coral reef growing around the top of an underwater volcano that emerged from the sea. It seems that no such coral island is to be found among the Aeolian Islands.

I was therefore very excited to see the corals growing on the old lava flow on the ocean bed when I visited the Hawaiian Islands in the Pacific. I went down in the Atlantis submarine from Big Island on 31 March 2008. The Atlantis submarine went down to a depth of 100 ft which allowed us to see the underwater life very clearly. I saw the beautiful marine landscape fashioned from corals and lava. It took twenty-five years for Hawaii's coral to develop on the lava flow and this process is helped by the cool water in the mid-Pacific.

I also had a spectacular view of underwater corals when we visited the Andaman and Nicobar Islands in India in 1982. The islands of Andaman and Nicobar consist of 572 islands, floating on the Bay of Bengal, over an area of 700 kilometres and 1,100 kilometres respectively, from the east coast of India near Myanmar (previously known as Burma), stretching up to the Philippines. One of the uninhabited islands is Barren Island, and this is the only island where an active volcano can be seen in India. The volcano erupted in 1991, in 1994–95 and in 2005, after remaining dormant for 177 years.

My visit to the Andaman Islands in January 2007 was unplanned and coincidental. In December 2006, I was in India for my younger daughter Tanya's wedding in Kolkata. She and her fiancé Andrew wanted to get married in the traditional way in India. When I was in Kolkata, I heard that the volcano on Barren Island had erupted again and that some geologists from the Geological Survey of India were planning to go there. I was interested in accompanying them and so I contacted Auditiya Bhattacharya, a geologist from the Geological Survey of India whom I knew. He introduced me to one of his colleagues, Pinaki Bandopadhyay, who was, at that time, going to the Andaman for a field study and was working on a research project with Mike Searle from Oxford.

On 3 January 2007, I arrived at Port Blair and met him at the airport. He took me to their camp office, where we discussed our programme for next five days. As I did not book a hotel in advance, I had some difficulty in finding suitable hotel accommodation but eventually found some before we left for Baratang in Middle Andaman, where several mud volcanoes are located. However, our first aim was to go to Barren Island, which was very difficult to get to, as there was no landing place; the last eruption had destroyed the traditional landing site. During that eruption, Pinaki was there and had collected some volcanic specimens and taken a video of the eruption, which he showed me. At present, no commercial or tourist boat goes there. To get there, we had to rely on the coastguard or an Indian navy vessel. Another option would have been to travel by helicopter. Last time, Pinaki and his colleague landed there with the help of the Indian navy.

With such a short time frame, we thought it would be difficult for us to arrange a trip by ourselves. So we decided in the first instance that we would go to the coastguard commander's office and discuss the whole issue. We met the deputy commandant. He confirmed that the volcano was active and this had been observed by coastguard personnel; they patrol the island and monitor the daily situation from the sea. However, at that time they were not able to provide us with any boat or take us to Barren Island. He did say that after 13 January an Indian naval ship would be able to take us, as four other scientists would also be going to Barren Island that week. Unfortunately, it was too late for me as I was flying back to Britain on 12 January. So, this time I had to abandon the Barren Island trip to carry out my visit to Baratang Island to study the mud volcano eruption.

We arranged a car and on the morning of 5 January started our journey from Port Blair, reaching Baratang Island by lunchtime. We drove along the Andaman Trunk Road, passed part of Jarawa tribal territory and took a ferry crossing before reaching Baratang Island. At Baratang, we went to the forest commissioner's office and met the newly posted young deputy director, who kindly arranged a guide and accommodation in part of their newly constructed forest guesthouse. The old portion was already occupied by Professor Deb and his wife, who were also visiting the island. Professor Mihir Deb is Professor of Geology and director of the School of Environmental Studies at the University of Delhi.

With a local guide, we went to visit the mud volcano. I saw a dome-shaped mud heap with an orifice from which liquid mud was coming out. The area was barricaded with bamboo structures and visitors were not allowed to go very near. Throughout the barricaded area, there were one large, one medium and three small to very small mud heaps. Outside the area, there were a couple of spots with no mud heap but where watery/muddy spillages were taking place. The rest of the area was occupied with mud, gravel, stone and rocks. Pinaki started to collect interesting geological rock samples and I went inside the barricaded structures to observe the mud eruption closely. The largest mud mound was one metre high and cone- shaped, and eruptions were taking place every sixteen seconds – on each occasion, mud sludge was ejected from its mouth.

The opening of the cone (its mouth) was more than twenty centimetres in width; I was not able to gauge the depth but I heard that it had been traced down to fifteen metres. We took three samples: one from inside the mouth, one from the spillage area and one sample from outside the mound. I was not able to collect the gas samples as I did not have the appropriate instruments with me, due to my visit being unplanned. However, the existence of colourless gases could not be ruled out, especially as some people noticed at times a sulphurous smell. The mud was viscous and grey-coloured.

When I touched the erupting mud sludge, I found the liquid mud was very cold, contrary to my expectation! My earlier conception of mud volcanoes was associated with a hot spring that produces boiling mud. However, the Baratang mud volcano is not a volcano of this type. Later on, I found out that the temperature of some of the erupted material or liquid mud from the mud volcano could be as low as freezing point. The mud volcano of Baratang Island is in such a group, although its mud was not as cold as freezing.

Mud volcanoes are more common in Asia than Europe and the highest numbers of mud volcanoes are located in Azerbaijan and the Caspian Sea area. My main concern was the emission of gas and how it would affect surrounding areas or villages, although I could not smell or observe any gas emission at the spot with the naked eye. The nearest village is two and a half kilometres away, where nearly a hundred people live and there are forty cattle. However, regular gas analysis and monitoring does help to prevent any catastrophe. I noticed something interesting about the eruption pattern: when I was at the site on the fifth afternoon, the eruptions were mostly every sixteen seconds, but when I visited on the next morning at 9 a.m. The eruptions were every five seconds. The differences in eruptive activity might be related to the tide. Moreover, ejected mud also contains rock fragments. Pinaki also collected some rock samples and the analysis will hopefully reveal some of the characteristics of the underlying strata.

We had dinner at the guesthouse. Professor Deb and his wife were there. We were discussing the volcano and Barren Island. The Professor was concerned about a supervolcanic eruption occurring; what would happen? It would not only affect the Andaman and Nicober chain, but also India, Bangladesh, Myanmar and Indonesia, resulting in a major disaster. It was unfortunate that I could not visit Barren Island at this time, but I look forward to a future visit.

However, to prevent volcanic eruptions from becoming volcanic disasters, the following measures are suggested:

a) Risk or hazard areas are to be appropriately identified.

b) Regular volcanic activities need to be measured and monitored.

c) An emergency plan should be developed.

The most famous violent eruption with heavy casualties took place in AD 79 at Vesuvius, Italy. The eruption killed at least 2,000 inhabitants. The great cities of Pompeii and Herculaneum were buried during the eruption of Mount Vesuvius. I visited the excavated ruins and saw the plaster casts of bodies, replicas of the postures in which those people died.

Scientists are still puzzled about the cause of death, as the bodies were covered with ash. Suffocation was therefore most probably the cause. During the eruption, the volcanic gases and ash with a temperature of 500°C formed an ash cloud, a pyroclastic flow which swept over the town, killing thousands of people.

The last eruption at Vesuvius took place in 1944 and since then Vesuvius has been quiet; nobody knows when the next eruption will be.

Near the Arctic Circle, surrounded by the Arctic and Atlantic Oceans, there lies Iceland, the land mass of 'fire, ice and water', created by submarine volcanic eruptions caused by the movements of plates only seventeen million years ago. In my childhood, I was very confused by the name of the country called 'Iceland'. My impression was of a land full of ice, and, secondly, was the pronunciation issue between 'Iceland' and 'Island'. This is because I was fairly unfamiliar with this politically isolated volcanic island up until January 1968, when I took up my first medical post in Scotland as a pre-registration house officer at Ballochmylle Hospital in Ayrshire. My predecessor was an Icelander who left before handing over the care of patients to me as a trainee house officer. Then, politically, Iceland came into the limelight in the 1970s when there were clashes between Icelandic gunboat's and British warships due to a dispute over offshore fishing rights.

I came into contact with some of the Icelanders in the 1990s when I was conducting a clinical trial on ICEROSS (Icelandic Roll on Silicone Socket) among amputees. Vascular disorders leading to amputation are not uncommon in Iceland and that is the likely reason why the ICEROSS prosthetics were invented in Iceland.

In 1996 in Iceland, there was the devastating Grímsvötn volcanic eruption: a four-kilometre-long flow with a ten-kilometre high steam cloud. It also caused a glacial flood, destroying bridges and land mass. I still remember the spectacular view of floodwater with icebergs hurtling across, which was shown on live television.

The enormous volume of water was a result of the ice in the subglacial lake melting and filling up the Grímsvötn caldera, because of the heat of the newly erupted volcano. I now understand why Iceland is called the 'land of ice, fire and water'.

I had always wanted to go to Iceland but many people discouraged me. This was because of the harsh climate, uninhabitable wilderness, short summers and dark, prolonged winters, expensive standard of living and lack of industry. However, Iceland's volcanic and geothermal features had always attracted me. There are glaciers, geysers, thermal springs, fumaroles, lava flows, mud pots, craters, calderas, igneous plugs and active and non-active volcanoes.

So, my next visit was naturally to Iceland, land of natural wonder, which I made in July 2005. This time, I accompanied a group of people who were geologists or interested in geology. The team was led by Alan Clewlow, an experienced geologist who knew Iceland very well. Since 1993, he had escorted a small group of people every year. This time, he was going to the interior of Iceland, where there was no access other than for two or three months in the summer, as, during the rest of the year, most of it was covered by snow and ice. So I thought that, for volcanic experiences, it would be ideal for me to accompany him on the interior tour of the geologically young country. The country is only sixteen million years old and lies on the Mid-Atlantic Ridge, where the North American and Eurasian plates meet. The two parts drift apart at a rate of two centimetres per year, cutting the island in two. We saw the Ridge in two places, one in Þingvellir National Park, the other in Grotagja, near Mývatn. In Þingvellir, we saw a series of deep rifts on the surface, from which lava poured. In Grotagja, we were able to locate a spot of iron bar on the ridge where the North American and Eurasian plates are separating. We measured the gap and it was fifty-five millimetres on 30 July 2005; in 1986, it was forty-five millimetres, hence a difference of ten millimetres over a period of nineteen years. I kept a record of it by taking a photograph at the time of our measurement on the gap of separation.

Alan Clewlow kindly took a photograph of me as I stood with one leg on the Eurasian and the other one on the North American plate.

During the plate separation, besides lava eruption, there might also be a low-magnitude earthquake. It seems that volcanic eruptions and earthquakes follow each other. The Mid-Atlantic Ridge is a 10,000-mile-crack in the ocean floor that created Iceland as a land mass between the submarine Reykjanes Ridge to the south-west and the Kolbeinsey Ridge to the north in the Arctic Ocean.

Some parts of Iceland emerged from the sea, the newest one being the island of Surtsey, located south of Iceland, which we saw from the air just before our flight landed in Reykjavík. In 1963, on the south-west corner of the Westman Islands, this volcanic island of Surtsey sprang up from the sea.

The Tjornes peninsula lies on the Tjornes fracture zone, connecting the northern volcanic zone that is Mývatn, Krafla in northern Iceland and the Kolbeinsey Ridge. The Tjornes peninsula is an elevated strip of land, which has been uplifted about 600 metres during the last million years. Geologically, it is made of thick sedimentary layers inter-bedded with thin lava sections. We visited one of the coastal sections of the Tjornes peninsula, near a small river north of Húsavík, where we saw marine sediments containing marine animal remains and fossils. I was able to collect part of the sediment, which was packed with Pleistocene fossils like bivalves and gastropods. I keep it at my home in England.

Húsavík, meaning 'House Bay', is the largest town in the north-east of Iceland and is Iceland's whale-watching capital. Húsavík is a natural harbour that was created as a result of a volcanic fault displacement that cuts through the town.

Although I am not very fond of whale-watching, I did not miss the opportunity of taking a whale-watching excursion from this harbour during my visit to Húsavík.

It was summer and the waters of Húsavík swarmed with whales, and we saw many minke whales and dolphins but not blue whales. I met a London university student, who was from near Húsavík and was doing research on whales. It seems that it is a good location for whale research.

Puffins are the most numerous birds in Iceland, which we saw in their thousands on a small, uninhabited island near the Arctic Circle on the same trip. Our trip was near the Arctic Circle, the weather was not too bad and it was not too cold; the sea was not rough and the sun was shining.

Mývatn means 'Midge Lake'. The lake is rightly named, as I realised when we were strolling round the lake after our evening meal. Although they are harmless, our walking became miserable with these tiny creatures constantly attacking our faces. The evening light showed on Lake Mývatn, and this was a sheet of red and green colour on the horizon. We were not sure whether this was a part of the Northern Lights phenomenon or a simple rainbow. It was summer; the Northern Lights (or Aurora Borealis) are best seen on cold, clear nights – mostly in the winter skies, though some say also in autumn and early spring. The sky over Lake Mývatn is one of the best places for such a light show in Iceland. The Northern Lights originate from the sun, and are created when solar particles collide with gas molecules in the Earth's atmosphere.

Lake Mývatn and its surrounding areas show the evidence of geological formation over various periods, and the landscape, which was moulded by glacier and volcanic activity, has constantly changed. No wonder we stayed four nights at the local hotel in the village of Reykjahild, north-east of Lake Mývatn! My room in the hotel faced directly towards the sun. In summer, visitors in Iceland experience 'white nights', when night never gets completely dark. I had experienced this on the first day when we arrived at Reykjavík, the capital of Iceland. I could not sleep well because of the white night and so, this time, I made sure that the room in the Reykjahild hotel was made properly dark by appropriately drawn curtains.

When I had a shower in the hotel room, I found that the water was very smelly, a kind of fishy or sulphurous smell. So, I thought it best to buy a bottle of water for drinking purposes, although I felt uneasy buying water in Iceland, as I knew that Icelandic water was supposed to be very good to drink. Then I discovered the other tap, which was drinking water. The shower water was geothermal hot water and that was the reason it was so smelly.

Next to the hotel was the white-coloured village church, Reykjahild Church. Local legend claims that, in August 1729, a remarkable thing happened here. During the period of 1724–29, ten kilometres north-east at Leirhnjúkur, there were earthquakes and volcanic activity originating from the Krafla fissure system, resulting in an outpouring of Leirhnjúkur lava. This was termed 'Mývatn fire'. Fast-flowing lava descended from the hills and stopped miraculously near the church, which deflected it from destroying the surrounding farmlands. Today, one can see the stretches of lava that avoided the white church.

'Krafla fires' took place between 1977 and 1984. We visited the eruption site and saw the recent lava flow. We visited Víti (Hell) Crater, climbed the crater and had a good view of the crater lake, the Krafla caldera. Geothermal activities were still ongoing, and nearby was the Krafla geothermal power station, where natural steam drives the turbines and generates geothermal power. Geothermal energy in Iceland is very widely used for heating and the production of electricity. The Krafla geothermal power station was built between 1975 and 1977, located at the ideal site for power generation. It produces sixty megawatts of electricity. However, since its opening in 1977, its production has been hampered due to various earthquakes and volcanic activities. The maintenance of the underground pipes is made difficult because of carbon dioxide and sulphur dioxide corrosion. We did not visit any other geothermal power stations, so I am not sure whether similar problems affect other plants or not.

Iceland has three major geothermal power plants and 17 per cent of the country's electricity is produced by them. The other two plants are the Svartsengi power plant and the Nesjavellir power plant, which meet 87 per cent of the heating and hot water requirements of the country. They are situated in the south-west and south of the country respectively. Over 90 per cent of houses in Iceland are heated by geothermal energy, which is the cheapest and cleanest form of energy, making the Icelanders proud to be the least polluted country on Earth.

The largest lava desert that I have ever seen was when I visited the Askja caldera. This is the part of the Askja fissure system, which is the longest in Iceland, at two hundred kilometres long and twenty kilometres wide. We saw three different layers of basalt lava flow on top of one another in the order of oldest, old and fairly new, formed due to the emptying of the magma chamber during the various Askja volcanic eruptions. The most recent eruption occurred in 1961.

There was very little vegetation and we could see distinct edges where the lava flow had stopped on each occasion of a volcanic eruption. Askja is situated in the interior of central Iceland, surrounded by the Dyngjufjöll mountain range, and our journey to that remote location was through some difficult tracks and spectacular scenery. We passed through gravel roads, narrow bridges, rivers, blind rises and bends, lava fields, sandur and the mountain scenery of Herðubreið, the highest table mountain found in Iceland. Sandur is a result of a vast sheet of ice melting onto land as lava flows erupt under the ice. In Iceland, we saw many table mountains. A table mountain is nothing but a flat- topped plateau that is formed by subglacial eruptions. On the top of the Herðubreið table mountain, at the peak, there is a cinder cone formation created when the eruption broke through the ice. Within the Askja caldera lies Iceland's deepest lake, with a depth of 220 metres. Lake Öskjuvatn was formed when the roof of Askja's magma chamber collapsed and it filled with groundwater. Next to it is another Víti Crater, filled with geothermal spring water. Both were created as a result of a large, explosive eruption in 1875. To reach there, it took over thirty minutes to walk along the path after driving through the ragged countryside. This walk was through a landscape I could only describe as lunar. To get into the Víti

Crater and Öskjuvatn Lake, we had to climb up and then down over the slippery slope.

Some of us took a quick dip in the sulphurous warm water of the Víti Crater. My journey back was slow and I enjoyed the solitary atmosphere while passing through the lunar landscape. I stopped and stood and took photographs.

Some of our group members passed by me and some stopped and walked with me. One of them informed me that astronauts had undergone some training on this surface before they landed on the moon. I am not sure how much truth there is in this, but it is certain that many space scientists expect to find similar types of ragged landscape on some of the other planets in the universe.

Iceland has many waterfalls and we visited two; one was in the south, called Gullfoss, and other one was in the north, called Dettifoss.

Gullfoss is on the Hvítá (or White) River and it is one of the largest waterfalls in terms of volume in Europe with a drop of thirty-two metres in two falls. This has produced a canyon seventy metres deep and two and a half kilometres long in post- glacial time. The main fall is over a basalt lava flow, which is harder than the sediment below.

Dettifoss is on the Jökulsá á Fjöllum river, which flows from the Vatnajökull glacier and is the most powerful waterfall in Europe, with a discharge of 200 to 1,500 cubic metres per second, a drop of forty-four metres and a hundred-metre-wide fall. Both have different characters but are impressive to look at.

'Geysir' (or 'geyser') means 'gusher' and the Great Geysir is one of the greatest attractions in Iceland. Other famous geysers of the world are Rotorua in New Zealand and Old Faithful in Yellowstone National Park, USA.

The Great Geysir, which started erupting in AD 1300, has not erupted fully since AD 1916. However, it does occasionally now, following an earthquake in 2000 which appears to have cleared internal passageways that had previously been blocked. Although we were not able to see the eruption of the Great Geysir, we were able to see another geyser one hundred metres south of the Great Geysir, which erupts every ten minutes to a height of twenty to thirty metres.This geyser is called

Strokkur, and it is nothing but a spout of white column of boiling water. It is no doubt a great spectacle to watch but the underlying mechanism of geyser formation is difficult to understand. It is a complex phenomenon, consisting of dissolved gases, temperature and the pressure of water vapour in the cavity above the water; I noticed that water was sucked into the cavity before eruption. It is a vent from which hot water and steam are violently and periodically ejected at the surface in a volcanic area, caused by the heating of groundwater by subsurface magma. A geyser is a high-temperature geothermal area with a base temperature of around 250°C. In the surrounding areas, we also saw fumaroles, mud pots, dormant geysers and hot springs. Iceland has got at least 250 geothermal areas and 780 hot springs.

Apart from the Great Geysir, we visited another three main geothermal areas; one in Hveravellir and two around the Mývatn area. All were located in the interior of Iceland and access was not always easy.

I have always been interested in fumaroles.

A fumarole is part of a volcanic vent from which volcanic gases escape. Some gases are toxic and some are not. I was patiently observing the emission of the fumaroles and the person who was standing next to me was Tom Sharpe. He is a geologist and works as a curator in a museum in south Wales. He asked, 'How toxic do you think the gases are?'

I answered, 'I am sure the gases are not so toxic and certainly are not as toxic as I have seen in the Aeolian Islands. Toxic emissions must be within permissible limits. It was interesting to see a rope barrier and wooden platform so that visitors have a good and direct view without exposure to toxic emissions. From a safety point of view, the visitors in Iceland are more protected from direct exposure to fumaroles gases than the visitors in the Aeolian Islands.'

We visited three geothermal pools and spa areas, the largest one being the Blue Lagoon, near Reykjavík, the smallest one in Hveravellir, and third one, the newest, Mývatn Nature Baths. We bathed in Iceland's newest geothermal spa, which opened in June 2004. I found that there were different temperatures in the various corners of the baths. In some areas it was too hot to stay long, and in other areas it was relaxing to dip in clouds of rising steam. It was enjoyable to swim in particular areas of the pool. The geothermal water of the pool is usually drawn from a depth of up to 2,500 metres.

The Blue Lagoon is more artificial than the natural bathing spot, created in the middle of the vast lava blocks and filled by the outflow of the Svartsengi thermal power station.

Hveravellir is a geothermal hot pool, where we stopped while crossing the central highlands through the difficult, bumpy tracks of barren landscape on the way to Akureyri from Gullfoss.

Travelling around Iceland and driving through the Icelandic country road and tracks is an unforgettable experience. A good and reliable driver familiar with local conditions is important. Every year, tragic deaths happen on the Icelandic roads. The road accident statistics showed that 131 people died in 104 accidents over a period of four years (2000–2004). Throughout the journey, our driver, who was a local man, did an excellent job. We ended our journey by giving him a token of our thanks.

Eyjafjallajokull, a 100 square kilometre (39 sq mile) glacier is located in south-east Iceland, near the Atlantic. The icecap of the glacier covers a volcano at the height of 1,666 metres (5,466 feet). In the past the volcano has erupted in the years 920, 1612 and 1821-1823.Recently,in the year 2010, it has erupted twice: on 20 March and 14 April. After a number of small earthquakes and some unusual seismic activities over two to three months, molten lava shot into the sky on 20 March. It was a spectacular sight and not dangerous. Many tourists flocked to the site for a glimpse. The eruption gradually slowed down until on 14 April, when Eyjafjallajokull erupted again beneath its icecap, causing melt water to flood and raising the surrounding rivers by ten feet, requiring 800 people to be evacuated. A plume of ash steadily rose into the atmosphere at a height of more than 20,000 feet and then continued to climb higher and higher. With the wind, it spread towards Britain and mainland Europe.

Following Met Office advice, British airspace was shut down from 15 April 2010 onwards. On 16 April I was supposed to travel Iceland to witness the latest volcanic eruption but I had to cancel my trip. For six days British aerospace was 'no fly zone' and all over the world millions of passengers were stranded as flights were cancelled in the UK, Ireland, and Northern Europe as well as some other airports in the rest of Europe. The freight transport of vegetables and flowers from Africa, South Asia and South East Asia for British supermarkets was also disrupted, resulting in shortages of certain items.

Many stranded passengers were rescued by rail, road, ferry and ships; although in some places passengers had to pay high prices for the fare for these alternative arrangements. The British Royal Navy's ships were also deployed in Europe to rescue military personnel as well as some civilian passengers.

The shut-down of the airspace was the precautionary measure taken by the air traffic officials to prevent high altitude disasters in case the air-planes were caught by the volcanic ash. Volcanic ash consists of tiny fragments of rock, glass and sand. The turbines suck in the ash, filling the engines and damaging the fan blades. The high temperatures cause glass particles to melt, coating the inside of the combustion chamber, resulting in the engines shutting down and the plane to plummet. This would have catastrophic consequences unless an emergency engine re-starts, which is only possible at lower altitudes.

The classical example of such incident was a BA Boeing 747 flight with 263 passengers which was caught in the ash plume from the erupting Mount Gallunggung volcano in Java, Indonesia. The pilot was Captain Eric Moody and he was flying from London to Auckland on June 1982. When four engines failed due to the ash, he glided the aircraft for fifteen minutes, dropping from 37,000 feet to 1200 feet before the engine restarted. The ash was on the windscreen and the visibility was almost nil but he was able to land in Jakarta safely.

At present there is no safe level of ash for aircraft to fly; further research is required and for the aircraft manufactures who must design aircraft and engines that can overcome ash-related aircraft failure.

In Britain, the ash cloud was so high up in the atmosphere that it was very difficult to see with the naked eye from the ground, although on the 15 April in the afternoon I noticed a cloud shadow on the horizon. By 18 April the Met Office reported that the volcanic ash had began to fall across Britain, coating areas in a fine layer of dust. The dust coating has been detected in the ground in the North, the Midlands and the

Thames Valley, as well as Heathrow and North Wales. In Iceland the volcanic ash fall was so thick that Icelandic farmers had to rescue the cattle from exposure to the ash.

The eruption also created a rare electrical storm, a thunderstorm that occurs when lightning is produced in a volcanic plume. In this phenomenon, electrical charges are generated when rock fragments, ash and ice particles collide in a volcanic plume.

The health agency which is monitoring the Icelandic volcanic eruption advised patients with existing respiratory disorder to stay indoors and keep their inhalers and medications at hand, although the present plume of volcanic ash was not considered to be a significant risk to the health. Volcanic ash and gas have the potential to kill people and, according to one report in 1783 on the Laki Iceland eruption, sulphur gas was responsible for killing a quarter of Iceland's population and 23,000 people in Britain.

Volcanoes are probably less harmful than earthquakes. Our understanding of volcanoes is getting better as the slow process allows scientists the opportunity of studying the behaviour more efficiently. Preventive measures, including early warning and evacuation, are possible nowadays in most cases. However, more emphasis is given to the long term or after-effects rather than acute episodes.

In the last century, 55,000 people died from volcanic eruptions throughout the world and some of the most violent eruptions that took place in recent memory were Mount St Helens in the USA (1980), Parícutin in Colombia (1952), Unzen in Japan (1990–95), and Mount Pinatubo in the Philippines (1991).

Scientists believe that the last major supervolcanic eruption occurred seventy thousand years ago at Yellowstone National Park, USA, and another major eruption like that is due now.

However, scientists are not able to predict when and where it will come. Hydrothermal activities are still ongoing at Yellowstone, including the 700 metres of Yellowstone Lake, and this attracts thousands of visitors every year.

The 1980 eruption of Mount St Helens, USA, produced an eruption twenty kilometres high into the atmosphere. Suffocation, thermal injury and trauma were the main causes of death from the eruption.

Volcanic eruptions not only destroy but also create new land masses. The classic example of this is Hawaii. To explore this phenomenon I joined Alan Clewlow's Volcanic Experiences tour. All together, there were sixteen people on this tour, one of whom was from Canada and the rest were all from Britain. After the eleven-hour flight from London on 24 March 2008, we arrived in Los Angeles. From there, it was another five-hour flight to Kona Airport in Big Island. When we arrived, it was late evening and I was surprised to see dimmed lighting at the airport and on the roads. I was told that the summit of Mauna Kea, the highest point on Big Island, is the site of world astronomical observatories, and that many countries have their own observatories situated here. To assist this astronomical observation at night, the lights are dimmed in the surrounding area. From the airport, a thirty- minute drive brought us to Keauhou Beach Resort Hotel where we were to stay for our eight days in Big Island.

Hawaii, the 50th state of the USA, is a land of volcanic origin. The Hawaiian archipelago consists of 132 islands, shoals and reefs spread over 2,451 km (1,523 miles) in the middle of the Pacific. Oahu, Kauai, Maui, Molokai, Lanai, Hawaii (Big Island), Niihau and Kahoolawe are the eight main islands. All of the Hawaiian Islands developed from a 'hot-spot' volcanic chain. Countless eruptions over millions of years formed shield volcanoes that eventually rose from the sea. Eruptions decrease in number and strength as a volcano moves away from the hot-spot and then gradually becomes extinct, but at the same time, a new volcano emerges and a further series of eruptions begin on the hot spot which is usually in fixed position.

Amongst all the Hawaiian Islands, Big Island is the only island where active volcanoes are still found and it is made up of five separate volcanoes. These are: Kilauea (most active), Mauna Loa (active), Mauna Kea (dormant), Hualalai (dormant) and Kohala (dormant). Kilauea is very active volcano which has been continuously erupting since 1983 and produces millions of cubic meters of lava each day. Most of this usually comes out of the fissures of the volcano and flows southwards down into the Pacific Ocean creating a new land mass and thus extending the area of Big Island.

On 26 March 2008, we went to Hawaii National Park to see the Kilauea volcano which was active in two places – on the summit and on the coast . We started our journey by coach from our hotel resort in the morning. We took Highway 11 and on the way we passed through coffee plantations, various resorts, macadamia nut plantations, sites of old sugarcane plantations, the Captain Cook monument, private ranch lands, areas of rainforest and several lava deserts created by Mauna Loa in the last century. The most recent one was created in 1950 when a village was destroyed by the lava flow.

As we were approaching the National park, we all got excited when we saw brown-coloured fumes erupting from the Halemaʻumaʻu crater of Kilauea at a distance. The sky was covered with the volcanic smog which was affecting not only Big Island but also other neighbouring islands. Of course this depended a lot upon the direction of the wind. Some of us also smelled the gases. When we arrived in the Park we were told that most of the 10.6 miles (17.1 km) of the crater rim was closed to the public, as Halemaʻumaʻu generates toxic gases, fumes, fine ash and lava, and from time to time there are tremors as well. Crater Rim Trail and drive between Kilauea Military Camp and Chain of Craters Road were closed. It meant that we were not able to go as close as we wished to, as flying near the plume was also banned due to the hazardous conditions. We noticed this when we had our helicopter tour over the island.

We therefore had to content ourselves with a view from a distance, but we managed to see the eruption of a plume of gas and fine ash particles rising 1.5 km above the crater. As we stayed overnight on the caldera area, we were able to see eruptions at different times of the day and night. With wind, the plume's direction changed from time to time. We saw that the colour of the plume was sometimes white and sometimes brown; we were told that the browner the plume was, the higher its ash content.

There was a brilliant glow of red, orange and yellow visible at sunset and early evening and at first we were uncertain as to whether this was the result of lava or underground magma, but were later informed that it was the latter.

Although Kilauea has been erupting continuously since 1983, a new vent appeared in Halema'uma'u recently, on 12 March 2008, releasing 2,500 metric tonnes of gas per day. The first explosion of this vent took place on 19 March and boulders were flying over an area of 75 acres. The explosion inside Halema'uma'u crater was the first since 1924. Local people said that the goddess Pele was clearing her throat. I gather that sulphur dioxide emission rates from the summit have risen to a hazardous level, so I went to the Kilauea Visitor Centre to find out further details, as the Hawaii Volcano Observatory and the Jaggar Museum were closed because of their proximity to the erupting crater. The Jaggar Museum usually gives a variety of information on Hawaiian volcanoes and records are also available, produced by the observatory. However, I talked to a park ranger who gave me some information on the typical content of magma gases, based on results produced using gas chromatography. Typically, gases are made up of largely water vapour (71%), but there is also carbon dioxide (15%), sulphur dioxide (5%), nitrogen (5%), sulphur trioxide (2%), carbon monoxide (0.4%), hydrogen (0.3%), argon (0.2%), sulphuric gas (0.1%) and chlorine (0.05%).

We stayed overnight at Volcano House, located near the visitor centres on the crater rim. Next day, we went to the coast to see the lava flow out to the ocean. We took Highway 11 (Hawaii Belt Road) again and then Highway 130 (Keaau-Panoa Road) up to Kalapana. We found some parts of the road were destroyed by the most recent lava flow. The town of Kalapana had already been totally destroyed by a 1990 lava flow. The Royal Gardens subdivision was also affected and eighty houses were destroyed. From Kalapana we walked down over the uneven terrain lava desert which included ropey (phaethon) lava and a blocky (aa) type of lava to reach the lava viewing sites, near Kapa'ahu. This eruption is from the cone of Pu'u O'o crater in the Eastern Rift Zone. Lava erupting from the flows travels 11 km to the sea through a tube system. The entry to the sea is an interesting sight as there is a huge cloud of steam, formed as the hot molten lava reacts to the cool sea water. We were able to see several hot, red lava flows in the midst of the sea waves, as well as the steam, though we were not allowed to go too close to the entry point. We also saw a new land mass on the coastline created by lava flow and already 230 hectares of new land has been built out into the sea.

When we were flown over the Eastern Rift Zone and coastal area by helicopter, we saw steamy, smoky red spots on the black landscape. The helicopter also flew over the beautiful rainforest region of Big Island but we saw some areas which were affected by acid rain. This was the result of greenhouse gases but these gases were not produced by factories, power plants or cars but by the toxicity of the fumes from Kilauea's volcanic eruptions. However, the vegetation grows very quickly and easily on Big Island and plant life begins to flourish as soon as lava cools down, taking only ten to fifteen years to fully develop and, in some of the very tropical areas, only five years.

This is contrary to what I saw in Iceland and in Sicily. If one compares how quickly the vegetation grows or life springs up in the lava deserts of these three countries, Hawaii will be at the top of the list with the fastest growth, followed by Sicily and then Iceland. I think that these differences are due to the variation of climate in the three countries, as Hawaii lies in a tropical rainforest region, Sicily in a sub-tropical region and Iceland near the Arctic.

After 500 years of being dormant, Mount Pinatubo in the Philippines erupted in 1991, resulting in hot blasts and producing thirty-five-kilometre-high ash clouds in the sky. Serious disaster or catastrophe was avoided by accurate warnings and the timely evacuation of 78,000 local, foreign and military personnel.

Due to limited resources, the effects of volcanic eruptions are found to be more disastrous in poor countries in comparison with developed countries. Such examples can be seen with the two volcanic eruptions in Ecuador (Tungurahua and Guagua Pichincha), where there were problems with the evacuation of large numbers of the affected population. The eruptions are heterogeneous and change with time. The fall of grey and white ash has not only affected the health of inhabitants in the form of eye, skin and respiratory disorders, but also contaminated the air, ground and water, which might be responsible for animal deaths as well.

Dry fog and toxic gases formed by the volcanic ash and aerosols are the most important health hazards for human beings.

The ash might be white or grey; the grey ash contains more metal or trace elements and this may affect the health to a greater extent.

The gas emissions in the atmosphere from volcanic eruption vary considerably between volcanoes. The analysis of some of the gases from volcanic eruptions show that carbon dioxide, sulphur dioxide, hydrogen sulphide, sulphur, hydrogen, hydrogen chloride, hydrogen fluoride, carbon monoxide and helium are present in the atmosphere. Ozone depletion and acid rain occur.

On the other hand, less harmful volcanic activity can be seen in the Taupo Volcanic Zone, New Zealand, which attracts lots of visitors. This volcanic eruption is not so destructive in nature in comparison with other volcanoes in the world.

The zone stretches in a line from White Island, north of the Bay of Plenty, through Rotorua and down to Tongiro National Park. Rotorua is called the Sulphur City and it survives due to a huge influx of tourists. Its thermal activity includes gurgling hot springs, bubbling mud pools, gushing geysers and typical sulphur smells.

The most interesting eruption of the Soufrière Hills volcano on the island of Montserrat in the West Indies started in July 1995, and activity is still ongoing. It created a new delta from the pyroclastic flows down the river where it entered into the sea. However, it is uncertain whether this delta will survive or ultimately be eroded by sea waves. The flows also produced convective clouds of ash, which fell over the island; the ash contained silica, which can affect the lungs of human beings.

Since the eruption, most of the original population of 12,000 have left the island altogether. This reminded me of a nursing colleague at Scottish Power, called Caroline Kitchen, who lived and worked in Montserrat before the eruption. She showed me some pictures and told me how the eruption affected the area where she used to live.

There have been concerns about the respiratory symptoms of the children living in the ash-fall areas, and a survey in February 1998 showed that volcanic ash emissions had adversely affected the respiratory health of Montserrat children.

By contrast, in a rich country like Japan, the children wear protective clothing around Mount Unzen, and volcanic bomb shelters have been built to provide shelters from volcanic bombs or blasts.

The future efforts of preventing volcanic eruptions depend upon prediction and damage limitation, for which appropriate funds and equipment are needed. However, sometimes scientists find it difficult to distinguish between active, dormant or extinct volcanoes, which might lead to disaster or vigorous outcomes. Scientists from the USA and Britain, in partnership with the Italian government, have devised a quantum-cascade laser that might detect impending eruptions, though this is, at present, in the experimental stage.

Measuring the plate-splitting of the Eurasian and North American plates, Iceland

CHAPTER THREE

Driest and Wettest

Death Valley, USA

Since childhood, I have thought of hot deserts as being the driest places on Earth and the wettest as being Cherrapunji in India. As I grew older, there have been certain changes in my ideas, so I recognise that the driest places on Earth are the valleys in the cold desert of Antarctica, where there has been no rainfall for the past 2,000 years.

However, Death Valley, in California, USA, is the hottest and driest place on earth, where the recorded rainfall was 0.1 inches (3.0 millimetres) and the temperature topped out at 134°F/57°C in 1913. A similar rainfall is noted in the Arica Desert in Chile, South America. The Gobi

Desert in central Asia and part of the Sahara Desert have an average recorded rainfall of 0.2 inches (five millimetres) and one inch (twenty-five millimetres) respectively.

It is not easy to visit all these places. However, when I visited California, in 1988, I stayed with one of my old friends, Dr Madan Mukherjee, who is a cancer specialist practising in Bakersfield, California. I flew from London to Los Angeles. His daughter Misti received me at the airport and drove me 120 miles to their home in Bakersfield. We arrived in the middle of the night. The roads were good and the sky was clear. She drove well. Madan and his wife Dolly took me to visit local places. Madan talked about the Grand Canyon, Death Valley National Park and the Mayan civilisation. Some of his visits, especially those concerning the Mayan civilisation, he wrote about in a popular Calcutta-based Bengali magazine. Although we were not very far from those wonder sites of the Earth, he stopped me visiting those places on that occasion and asked me to programme it for next time. Of course, I had very few days to spare. I knew I would not be able to make it this time. Instead, sitting in their backyard in the evening, I enjoyed the wonderful view of the Californian desert.

Nearly eighteen years later, I visited Death Valley National Park in March 2006. This time I flew from London to San Francisco. After spending a couple of days in San Francisco, my journey for Death Valley began. We followed some of the Pioneer or Native American trails but we drove through well-paved, mostly highway roads in air-conditioned coaches. On the way, we passed through the Napa Valley, Sacramento, Lake Tahoe, Carson City, Virginia City, and visited the Yosemite National Park. We crossed the Nevada–California border at Lake Tahoe. Our last overnight stop was Tonopah, Nevada, before entering Death Valley.

Death Valley is located in the Mojave Desert, California, in between two mountain ranges, the Armargosa range and the Panamint range. Geologically, Death Valley is a subsided basin floor, leaving precipitous mountains on all sides, which has created a blocked land of 5,000 square miles. The lowest place is Badwater, 282 feet below sea level, and the highest is Telescope Peak, 11,049 feet above sea level. I thought the place was called 'Death Valley' because this is an arid area where nothing grows, but the name comes from American history when the American pioneers and settlers were looking for gold in the California mountains. In 1849, some of them perished along the way when they were forced to cross the burning sands to avoid the snowstorms of the Sierra Nevada. In 1861, the place was formally named Death Valley, as those pioneers faced extreme hardship in brutal and hostile environments.

The highest temperature in summer is usually around 115°F/46°C and the lowest is 65°F/18°C, usually in December or January. The highest recorded temperature was 134°F/57°C in July 1913 and the lowest was 15°F/-9°C in January 1913. That year was also the wettest year with 4.54 inches of rain, although the average annual rainfall in Death Valley is not more than two inches.

The day I visited was 21 March 2006, and the temperature on that day was 75°F/24°C. Before entering Death Valley – the night before – we stayed in the historic silver mining town of Tonopah. It was a cold night and there was heavy snowfall. In the morning, we drove south and crossed the state border to enter Death Valley. The roads were mostly highways except for some side roads, which were not bad at all. In the desert, I had not expected such nice roads. Of course, we stayed on major paved roads or highways, as I understood there was a network of roads paved, unpaved and primitive, which provided access to all corners of Death Valley. I saw very few gas (petrol) stations or medical facilities. Moreover, on the highway we were not able to stop even to take photos, as there were hardly any lay-bys.

We passed through snow, sand, mud and rocks, canyons, volcanic craters and white salt. Snow was on the mountains, and rocks appeared in various types and colours, including black, brown, orange and green. The desert appeared to me more semiarid than arid.

It seems that there is rainfall, but it is minimal and mostly desert plants grow and survive in this scant rainfall. I saw creosote brush, desert holly, palm trees and Joshua trees.

I was interested in collecting weather information for the day I visited Death Valley. I went to the visitor centre, which is located in Furnace Creek, an oasis in Death Valley. It is situated 190 feet below sea level. Inside, there is a museum, souvenir shop, bookshop and park headquarters. The museum tells the story of Death Valley, which is fascinating. From the rangers, I found out that the temperature over the last twenty-four hours had been between 24°C/75°F (high) and 11°C/52°F (low). This was the morning report of Tuesday 21 March 2006.

In the nineteenth century, there was a mining boom in Death Valley. Minerals like silver, gold, borax, talc, salt, gypsum, celestite (strontium sulphate), sodium and potassium nitrate, and sulphur were discovered on the floor of Death Valley and its surrounding areas. On the floor of Death Valley, I noticed 'cotton ball' crystals, which are fluffy concentrated borax salts, formed as a result of evaporation.

Workers used to collect the 'cotton ball' crystals, which were then processed to make borax. I visited one of the old refinery ruins of the 1880s, Harmony Borax Works, located north of the Furnace Creek visitor centre. I saw disused wagons that were hauled by eighteen mules and two horses from Harmony Borax Works for 165 miles to Death Valley junction, a railway station now abandoned. A twenty-mule team hauled two wagons and a 1,200-gallon water tank through the Mojave Desert. Each wagon carried twelve tons of borax.

Badwater is the most famous place in Death Valley and it is only 17.6 miles from the Furnace Creek visitor centre. We drove southward along the main driving route of Death Valley and, on the way; we passed through beautiful scenery of geological rock formations, mostly barren rocks.

As soon as we reached Badwater, we saw a basin that looked like a white salt pan. There was a small pool of water, mud, sand and gravel on the basin floor, but it disappeared as we walked through; I could see from the miles of white salt flats that it was once a lake. If time permits, then one can easily cross the salt pan from one end to the other. Once it was a deep, vast fresh water lake but thousands of years of dry, hot weather and wind have been responsible for the evaporation of the fresh water, leaving only the salt flats.

There is no outlet; moreover, the snow water and rainwater from the surrounding hills flow, but dry up before they reach the basin floor. I was observing the vastness of the barren and lifeless salt pan when I suddenly realised that I was standing on the lowest point in the western hemisphere, 282 feet below sea level. I tested the salt from the salt pan and collected a crystal sample as part of my souvenirs.

After visiting Badwater, our journey continued through the Amargosa Valley, past Spring Mountain, towards the world- famous entertainment city, Las Vegas. I stayed three days in Las Vegas and visited the Grand Canyon before flying back to London.

Badwater reminded me of my visit to one of the other driest places on Earth in 2002. That was the Sahara Desert of Africa, and this I did when my wife and I visited Tunisia.

Not very far from the Algerian border in southern Tunisia, there are closed depressions or basins, known as 'chotts'. We visited a couple of such desert basins, where we found heavily salted water and also saw crystals of white salt that were formed as a result of water evaporation from the basins in the hot desert climate.

The largest desert on earth, the Sahara covers most of Algeria, Libya, Egypt, Mauritania, Niger and parts of Morocco, Tunisia, Mali, Senegal, Chad and Sudan. The desert area ranges from 8,633,000 to 9,982,000 square kilometres on the basis of the expansion and contraction of the Sahara Desert region. The movements are southward or northward (mainly southward). The alleged southward expansion of Saharan Africa is due to droughts and land mismanagement, like overgrazing, increased cultivation and cutting of firewood. The northward expansion I saw when we visited south-west of Tunisia, near the Algeria border and far away from the capital city of Tunis on the Mediterranean coast. It was on our camel safari through the heart of the Sahara. We found a village that had been abandoned due to the Sahara Desert engulfing it.

The average rainfall in southern and central Tunisia is less than 200 millimetres. However, in one place in the Sahara Desert, which we visited in southern Tunisia, I was told there had been no rainfall for a number of years. No wonder the Sahara is expanding.

Mount Waialeale in Hawaii is the second wettest place on Earth, with an average rainfall of 460 inches and Cherrapunji, India, is the first, with an average rainfall of 10,874 millimetres (428 inches) – its highest recorded rainfalls were more than 11,836 millimetres/466 inches of rain in July 1891. Mawsynram is three kilometres from Cherrapunji and recently took the title of the wettest place on Earth from Cherrapunji. Mawsynram has 11,873 millimetres (467.5 inches) of rain per annum.

I visited Cherrapunji twice, the first time in 1967 and then in 1980. The 1980 trip was a very short visit with my wife. In 1967, I was touring Assam along with one of my friends, Samir Sengupta. Samir is now an advocate practising at Kolkata High Court. From Calcutta, we went by train to Guahati, which is the state capital of modern Assam. From Guahati, we took a car and reached Shillong, the state capital of Meghalaya. On the way, we stopped at the local roadside market where we saw Khasi girls with their traditional customs and costumes, selling fruit and vegetables.

Both Cherrapunji and Mawsynram are part of the state of Meghalaya. The state was originally part of Assam and, since 1972, it has been one of the independent states of India. It consists of three hill ranges: the Khasi Hills, Garo Hills and the Jayntia Hills. Meghalaya means a place in the midst 'of the cloud', and the name is rightfully applied, as most of the time the sky is cloudy, foggy or misty. One third of the state is forested. Limestone caves, waterfalls and lakes attract many visitors.

From Shillong, we went on to Cherrapunji, fifty-eight kilometres south of Shillong. Cherrapunji is situated at 1290 metres above sea level on the Khasi–Jayntia hill range. It is the monsoon weather that is responsible for the torrential rains. Although the monsoon is beautiful in some parts of India, the people living in Cherrapunji are often fed up with the continuous rain, stuck at home with nothing to do. This is especially true of the people from other regions posted there, who are not accustomed to this type of monsoon weather. Gambling and drinking are not uncommon. They are also quite prevalent among the indigenous population, and I saw that the local people played and gambled with a darts-like game. The local tribe is mostly Mongoloid ethnically, and some say that they might be closer to Indonesians.

The monsoon in Cherrapunji lasts for four months (June to September), resulting in heavy downpours, plenty of waterfalls, mountain springs and green forests. The rains run off the mountains into the valley below. I remember, when I visited Shillong in 1967, a local female journalist took us to a scenic spot on the Shillong–Cherrapunji road where two beautiful waterfalls were to be seen. She spoke of these two waterfalls as two sisters. They jumped from the mountain into the valley and there was a legend behind this, but I cannot remember the details of her story because it was forty-three years ago.

The monsoon is beautiful in India. Artists, poets, writers and musicians from times past have described the beauty of the monsoon in different forms. I have also enjoyed its beauty in the hills, plains, forests, rivers and seas of India, especially the rain pouring in with storms, thunder and lightning after a long, hot, humid summer.

I remember how important it was to catch thunderstorms and lightning in a motion picture. Mr Satyajit Ray (1921–92), the great Indian Oscar-winning film director who is famous in art houses all over the world, was the first Indian film-maker to bring art into the Indian cinema. This is clearly visible in his great Bengali art movie *Pather Panchali*. Its beauty was in its simple village background and, to catch these various scenic views, the effort Ray went to was great. No wonder he won several international awards for this, as well as other art films he produced throughout his career. The shooting of *Pather* Panchali took place in a small village outside Calcutta, and Ray was trying to shoot the beauty of the storm: thunder and lightning, followed by rains. He tried three times to catch this before he eventfully succeeded.

I was nearly in my teens and one of my friends whose family was involved in the making of this film told me that the crew and artists ran to the location of shooting, sixteen kilometres from Calcutta, after hearing that the pre-monsoon cloud was gathering in the sky and there would be a thunderstorm; but when they arrived, that thunderstorm was over. In those days, that was a quite a distance and it was not easy to catch the storm, thunder, lightning and monsoon all in one shot, after collecting all the artists and other technical people, and reaching the location in time. However, Ray was lucky and was finally successful on the third attempt.

During the summer months, the Indian subcontinent heats up, generating a seasonal continental region of low pressure. Airflows reverse and the wind blows south-westerly across the Indian Ocean, accumulating considerable moisture that is deposited as heavy rainfall during the wet season from May to September. During the monsoon, the water vapour that is carried upward and northward hits the Himalayas, and rain falls. Scientists have linked the monsoon wind phenomenon over India to the uplift of the Himalayas and the Tibetan plateau, when India collided into the Asian continent, twenty million years ago. Winds flow from areas of high pressure to areas of low pressure. Water vapour is carried upwards in rising air and condenses to form clouds and rain. Every day, the sun evaporates water from land and sea. This stays in the air as water vapour, which is not visible until it cools and condenses, forming tiny droplets of water, which can be seen as mist, fog or clouds.

In Cherrapunji, the seasonal winds in monsoon times bring torrential rains that last for four months (June to September). For the rest of the year, the winds change direction, and over the next few months, hardly any rain falls.

The rain gauge is one of the simplest instruments used to measure rainfall and is made of a funnel leading to a graduated tube. The amount of rain that has fallen is measured every twenty-four hours.

In Cherrapunji, in 1967, I saw the rain gauge used but, as far as I remember, the method of rain collection was primitive in comparison with today's standard. The rain was collected in a jar and then the amount of rainfall was measured after pouring the collected rain into the rain gauge. I thought the method was laborious and there might be some pitfall in measuring rainfall by this method. Nowadays, there are other types of hydrometeorological instruments, which are easily available and give more accurate results. I believe that Cherrapunji is using better types of instrument nowadays.

During the British Raj, Cherrapunji was one of the last British posting stations on the eastern frontiers. Some British people served there as civilians or as military personnel. They liked this mountainous, picturesque place, which reminded them of their home, and the summer capital of Assam that is Shillong was not far from there. However, the monsoon weather and tropical diseases must have taken their toll, and nearby there is a cemetery where one can see where some of the serving men in British India were buried.

When I was working in Scotland in 1968, I met a doctor who came from the mission hospital in Cherrapunji. He was spending a couple of weeks in Scotland before he attended the diploma course in tropical medicine at Liverpool University. This indicated that, to work in India, especially in certain regions, one should have proper knowledge of tropical medicine. That inspired me to take the tropical medicine diploma at Liverpool University in later years.

Cherrapunji has changed ecologically from a small village to having many more inhabitants. The weather and human factors are responsible for this change. A monsoon brings heavy rains for four months and very little rain for the rest of the year. This brings problems with growing food; deforestation has also affected Cherrapunji. There are problems with obtaining clean or fresh water during the dry season.

It reminds me of a quotation from a famous poem, entitled *The Rime of the Ancient Mariner*: 'Water, water, everywhere, nor any drop to drink.'

The entire north-east region has one of the highest rainfalls in India. Cherrapunji and Mawsynram, on the southern slopes of the Meghalaya hills, record the highest rainfall in the world. The record goes back to 1850 in Cherrapunji and, in Mawsynram, to 1940. I gather that rainfall data in Mawsynram is not available for certain periods, because the rain gauge station was dysfunctional for many years and measurements restarted only in November 1996.

I asked myself, 'Why does the region have the highest rainfall in the world?'

The most logical explanation is that geographical and orographic factors are responsible for such variations in the monsoon rains in India. The region with the highest rainfall is surrounded by the Himalayas in the north and Arakan, Mizoram, Manipur, Naga and Patkai Hills in the east. These, and the presence of Brahmaputra, the Barak Valley system and the plains of Bangladesh and West Bengal, play a significant role in the development of surface pressure and lower and upper troposphere flow patterns over the region. Moreover, southerly moisture flows from the Bay of Bengal; the east-west orientation of the Meghalaya hills and the heights of Cherrapunji and Mawsynram are important additional factors for heavy rainfall.

Information on annual rainfall reveals that 70 per cent of rainfall occurs during the monsoon season (June to September). In the pre-monsoon season (March to May), the figure is about 20 per cent. The post-monsoon season is from October to December, when the rainfall is about 7 per cent. In January and February (winter months), the rainfall is less than 1 per cent.

Post-monsoon season reminds me of a childhood incident in Calcutta.

At the end of school, I was returning home through a playing field which was not my normal route. I was at that time ten or eleven years old. At the edge of the field, there was a muddy pathway and very few people used to use it. It was a low-lying area and for most of the monsoon season it was covered with rainwater. In October, it started to dry up but the narrow, muddy pathway was still wet, which made it a good harbouring place for snakes. This I had not realised until I got bitten by a water snake. It was nearly evening, getting dark; I did not know that a snake was lying on the pathway and accidentally I stamped on it. I was wearing shorts, shoes and socks. I was bitten above the sock line, near the calf muscle on my right leg. I gather that is the most common site for snake bites. However, I went home. My wound was dressed with some antiseptic and, to prevent the poison spreading, a knot was tied above the wound. I was not shocked but anxious. My mother took me to our distant relative-cumfamily doctor, who cauterised the bitten area. I described the incident to the doctor and the type of snake was a water snake, most probably non-venomous. I was kept under observation at home, but recovered quickly without any complications.

Snake bite is one of the major causes of death in India and, every year, more than 10,000 people die from snake bites. Snakes are found throughout the world, except Antarctica, and worldwide incidences of snake bites could exceed five million. 2,500– 3,000 species of snakes are distributed throughout the world, and about 500 are venomous. In India, about 250 varieties of snakes are found, of which about fifty are venomous. The most deadly one is the king cobra, and a single bite can produce enough venom to kill up to twenty people. Due to deforestation, urbanisation and ecology change, its population is coming down, however. The snake has a great mythological significance, and temple walls throughout India are decorated with snake sculpture. Snake charmers and snake worship are common in India. The king cobra is worshiped during the festival of Nag Panchami, and is garlanded around the neck of Lord Shiva, the Hindu god. Venom from the cobra is also used for the preparation of antivenom, which are effective in the treatment of bites from some of the poisonous snakes of the world. Australia is a pioneer in the treatment and management of snake bites. In Australia, about 3,000 snake bites occur per year, of which 200 to 500 require antivenom treatment. On the other hand, in the United States each year, 10,000 to 20,000 snake bites are reported, but fatal cases are very few.

Another of the wettest places on the Earth is Mount Cameroon, in Africa, where average annual rainfall is 10,160 millimetres (400 inches).

Steadily expanding Sahara Desert, Africa.

Chapter Four

Floods and Dams

Flood and pollution, Ganges

In geography lessons at school, one learns why Egypt is called the Gift of the Nile and on the other hand why the Yellow River (Huanghe) in China is called the River of Sorrow. In religious class, one is taught about Noah's flood and the Ark.

There are many legends about floods, especially in the ancient world, as civilisations grew on the banks of great rivers. Since ancient times, it was believed that because the human race could not control nature, then it must be an Act of God.

Christians believe that Noah's flood and the plagues of Egypt were Acts of God.

Noah and his family survived the Great Flood by building a ship (the Ark) by God's order. When the water receded, the Ark landed on Mount Ararat in eastern Turkey, supposedly in 2348 BC. Various expeditions in search of Noah's Ark are still ongoing, although some scientists believed that they have located the wreck on the seabed, near Turkey.

Similar stories come from other parts of the world, especially from China, America and India. There is a Chinese legend of an extensive flood that revolves around the goddess Nekua, who ultimately ended the flood by patching up the blue sky with five coloured stones. An Indian myth from the sixth century BC tells the story of Manu, who was advised by a fish to build a ship to escape from the coming flood. When the flood came, the fish towed the ship to a mountaintop.

Anthropological excavation in Iraq has discovered a layer of clay that supports the existence of an ancient flood. A great worldwide flood is recorded as an epic story, the Epic of Gilgamesh, on a Babylonian clay tablet, which was written 2,500 years ago.

Whatever the Biblical or mythological story, many scientists believe that, between 12,000 and 10,000 years ago, there was a universal flood due to a rise in sea levels resulting from the melting down of the icecaps. However, in this modern age, universal climate change (or global warming) has already resulted in increased coastal and inland flooding. Scientists have predicted that if global warming continues at this rate, then there will be a loss of land mass due to submersion.

This will be a disaster for people living near the sea or living on low-lying deltas. One hundred million people in the world live in low-lying areas like the Maldives, the Ganges-Brahmaputra Delta in Bangladesh, the Nile Delta in Egypt and the Mississippi Delta in the USA.

Melting glaciers appear to be the first victims of global warming. Its first human casualties in this century are the residents of a small island in the Arctic that is disappearing into the sea. This small island is in Alaska, and is named Susan Rock. The islanders have already become refugees and have had to migrate to the mainland.

In some places on Earth, flooding is a routine phenomenon, though the types of flood vary.

There are four types of floods:

1. Flash floods are the most dangerous type of flood, which appear and move quickly without any warning. Waters move at very fast speeds and rising water overflows its normal pathway. Their powers of destruction are unimaginable. The fast-moving water destroys buildings, obliterates roads and bridges, uproots trees and moves boulders or anything in its path. It carries a huge amount of debris, and walls of water can even reach twenty feet high. Flash floods commonly happen as a result of heavy rainfall concentrated over one area by slow-moving thunderstorms, or thunderstorms that repeatedly move over the same area, or heavy rains from hurricanes and tropical storms. Dam failures are sometimes responsible for flash floods, as a huge amount of water suddenly flows downstream, destroying any obstruction in its pathway. Another type of flash flood occurs in desert (arid) regions, in arroyos. An arroyo is a water-carved gully or a normally dry creek, found in arid or desert area, resulting from storms and rainwater that cut into the dry, dusty soil, creating a small, fast-moving river. Flash floods also occur suddenly near melting glaciers and snowmelt Mountains.
2. Coastal floods occur not only from heavy rains, but also as a result of high sea waves or ocean water flooding the coastal land mass, destroying houses, towns, sea resorts, and roads, bridges, fishing villages and farming lands. This type of disaster usually occurs because of storms or strong winds in coastal areas in the form of cyclones, hurricanes, typhoons or tsunamis created by earthquakes or volcanoes in the ocean.
3. River floods are the most common type, where flooding occurs along rivers. They are usually seasonal. They occur when winter snow melts, excessive rain falls or, as in monsoon

rains, when river basins fill too quickly and overflow their banks.

4. Urban floods occur when the rainwater cannot be absorbed into the ground. It occurs in under- or badly developed or lowland areas. The ground loses its ability to absorb rainfall and water runs off, overflowing drains, flooding roads, fields, basements, businesses and other premises.

The worst flash floods for fifty years in Britain occurred on 17 August 2004, when a ten-foot wall of water tore through the Cornish village of Boscastle, destroying buildings, the sixteenth- century harbour and a stone bridge, and sweeping cars out to sea. Boscastle, a slate stone village on a steep hill that converges down three steep channels into a single harbour, is a funnel for three rivers; heavy rain in one afternoon was too much for the narrow banks. Localised low cumulonimbus cloud formation around Boscastle was responsible for the heavy downpour, and between 11 a.m. and 6 p.m. 133 millimetres of rain fell.

A pretty Cornish tourist spot became a disaster zone within five hours, and both villagers and holidaymakers were trapped. It was a miracle that nobody was killed and only rapid rescue prevented a human calamity. 120 people were rescued from the rooftops and trees in four hours by seven helicopters.

In Britain, I live in a small village near the Meon Valley in rural southern England. The Meon is a small river that originates in east Hampshire and runs through one of the most beautiful valleys in the country. It joins the Solent, which leads to the English Channel. At least three to four times a week, I used to travel to my workplaces in Basingstoke as well as in Winchester and thus I crossed the River Meon several times from its source to its flow downstream. On my way, I saw the river appear and disappear in the midst of fields, woods, hills and meadows. The river, which originates as a stream, becomes wider as it runs downstream. I have seen the beauty of this river all year round. In summer, it dries up and, in late spring and early summer, the river becomes thin and narrow. When autumn comes, the leaves start falling; gradually winter approaches, and the River Meon starts to swell and the riverbanks burst or flash, resulting in groundwater flooding.

In the autumn and winter of 2000, there were the worst floods for forty years in southern England, affecting Kent, Sussex, Hampshire and the Isle of Wight. Many areas were cut off. I saw car and lorry drivers who had to abandon their vehicles or were waiting for rescue. Once, I thought my car would be stuck on the flooded road; however, I was able to escape without damage to the engine. My car was damaged on a couple of other occasions, though. Once, the left side of the front bumper was damaged, having struck a fallen tree, and, on another occasion, roaming pheasants were caught by my car when on the road trying to escape from the floodwater. Pheasants killed by road traffic are not uncommon. This I noted mostly on dark winter days on the road passing through the forests and woods, and I had to stop on quite a few occasions, seeing injured pheasants. I wish I could write a country diary, as I feel that more animals are killed on roads than by game shooting. Animal lovers would not like it.

In some places, there was muddy floodwater and sewage. The traffic was diverted. Railway services were affected and some motorways were badly hit. Some people lost their cars. A high street was caught in the torrent and a supermarket in one place caved in under the pressure of the water, and stock floated off into the street. There were problems with pumping out the floodwater from affected properties and keeping the swollen rivers flowing. The emergency services and the Environment Agency were stretched. Our front garden was damaged due to a burst pipe. Properties damaged by gales, burst pipes and floods were not uncommon.

In Hampshire, 35,000 houses on the flood plains of the Itchen, Test, Avon, Meon and the coastal areas of the Solent were vulnerable to flood; the flooding in the winter of 2000 cost the insurance industry over £600 million nationwide.

Floods in summer are very rare in Britain, but it did happen in 2007. Torrential rain brought extensive flooding throughout the country in June and July 2007. The most affected areas were Yorkshire, the Midlands, Gloucestershire, Worcestershire, Oxfordshire, Berkshire, south Wales and Northern Ireland. Flooding affected millions of people and took the lives of eleven people. Thousands of homes, businesses and crops were damaged, and damages were estimated to be over £2 million.

I can recall that, on 28 June 2007, I had to be in Sheffield and Leeds, but couldn't go due to the train routes being flooded. However, I was able to go to Leeds the next day on a different train route. On the way, I saw flooded areas near Doncaster, which looked like waterlogged paddy fields. The River Don flows through Sheffield, Rotherham and Doncaster, and it burst its banks, causing widespread flooding. This situation leads to many people suffering from stress. When I was working in a clinic in Yorkshire, I met someone who had flood-related stress. I also heard about a person who had shot himself because he was unable to cope with the effects of the flood.

All of Sheffield city centre was flooded, and even the out-of town shopping centre, Meadow Hall, was affected. When I visited Meadow Hall five or six weeks later, I found that some parts of the shopping complex were still closed for business, and work on the flood-damaged areas was ongoing. The nearby sixty-five-acre engineering complex of the famous steel company, Sheffield Forgemasters, was also flooded with a wall of water from the River Don. In Sheffield itself, two people died and many were trapped in flooded offices, buildings and cars. In many places, the floodwater was contaminated with sewage and, for some people, gas, electricity and water supplies were cut off. The flood defence scheme did not work in many places, and a lot of people had to be evacuated from their homes. June was one of the wettest months on record. The average rainfall across England was 14 centimetres (5.5 inches), which is double the average for June. Global warming has been blamed for this unusual event.

Similarly the November 2009 was the wettest month with an average rainfall of 217.4 millimetres. This exceeds all previous records since the meteorological records began in Britain. This heavy rainfall on 18 and 19 November led to severe flooding at Cockermouth, Workington and Keswick in Cumbria. Seathwaite in the Lake District (Cumbria) is known as wettest place in Britain and here the rainfall within twenty-four hours was 316 millimetres.

The rivers in Cumbria were swollen by this torrential rains running off the hills. The most affected rivers were the River Cocker and River

Derwent, which bust in certain places. Six bridges completely collapsed and one bridge was damaged though sixteen bridges, twenty-five roads and eighteen schools were closed at that time because of the high level of water. The most tragic incident was the death of a policeman who was engulfed by the flood water and washed away, as the Northside Bridge over swollen River Derwent in Workington suddenly collapsed, while he was on duty helping drivers and pedestrians to cross the old stone bridge.

The town Cockermouth which lies at the confluence of river Cocker and River Derwent was damaged most as a result of this flash flood. Torrents swept through the main street, damaging the shops, houses, hotels, churches, historical building, cars and various utilities. There was a large number of debris in the water. Many people were trapped in the flood water. At least 50 people were airlifted by RAF helicopter and 20 people were rescued by lifeboat crews but volunteer rescuers were also there. A total of 200 persons were rescued. Twelve hundred people in Cockermouth were left without electricity.

When I visited the place on 26 December 2009, nearly five weeks after the flood, I found that many shops, hotels and business premises were still closed because of repair work. I saw that giant drying machines were still active in some premises. On some areas of the river banks, I saw sand bags and flood sacs which were still there to prevent flood water from getting into the premises. Debris was also trapped in certain places. One or two local people pointed out to me some of the worse affected areas. Cockermouth's most famous building,' Wordsworth House', was also closed for Public, though this seventeenth century buildings and it contents were safe. This is the historical building where the poet Wordsworth was born in the year 1770.The back garden leads to directly on the River Derwent where poet used to swim as a boy. I tried to access the river front through its side road but I could not as the flood-related temporary fence for repair work was still in situ. Two temporary GP surgeries with one counselling room have been set up in flood-hit Cockermouth. The rebuilding of bridges and repair roads were ongoing and the final coast of repairing of roads and bridges was difficult to predict but it will certainly coast the Country Council millions of pounds. A new foot bridge in Workington, connecting the north and south, was urgently built by army. According to the British Insurance

Association the floods will coast them up to 100 million pounds. In Britain over three hundred major floods have occurred in ten years and no doubt these had a great impact on the country's economic.

The River Nile in Africa is the world's longest river (6,670 kilometres) and floods every year between June and October. When it floods, it provides the entire Nile Delta with water and fertile soil. The Egyptians cultivate crops and fruits. Without the Nile, Egypt could have been engulfed by the Sahara Desert and there would not have been an Egyptian civilisation. The Nile is Egypt's lifeblood and, since ancient times, everything has depended on this river, from food to river transport. The river flows from the south (the Nile Valley) to the north (the Nile Delta), connecting both upper and lower Egypt. In 1998, my wife and I, and three other people from England, spent seven days on a ship cruising on the Nile. We enjoyed the beauty of the Nile during the rising and setting of the sun; we passed from upper to lower Egypt, visiting temples and the Valley of the Kings.

The places of the rising (east) and the setting (west) suns were very significant for the Egyptian culture. The east was associated with the living and the west with the dead. So the pyramids and tombs were located on the west bank of the Nile.

Sometimes floods have caused destruction and disaster in the villages and settlements along the Nile river bank. On the other hand, there have also been periods of drought, resulting in disaster and famine. The building of the Aswan Dam is Egypt's great effort to control the mighty river.

Aswan (Assuan), the ancient Syene, lies on the right bank of the river Nile, 886 kilometres from Cairo. This is where Egypt ends and Nubia begins. The first dam at Aswan was built in 1902 and the dam was raised in height twice, in 1912 and 1934. After becoming president, G A Nasser (1918–70) decided to build a new, higher dam, beginning in 1959. To build the high dam, unless appropriate measures were taken, many of the Nubian temples would have been lost for ever. UNESCO responded to the appeal to rescue this ancient heritage, and fifty countries helped with money and expertise. The monuments were taken down or cut free from the rock into which they had been built and reassembled in areas safe from rising waters. Those which could not be moved stayed below the water. Many villages were drowned and the population moved to a new site. The construction of the dam was completed in 1971. The reservoir is now known as Lake Nasser (600 kilometres long and fifty kilometres wide) and stretches 510 kilometres (317 miles) to the south, with a capacity of 150 billion cubic metres (5,300 billion cubic feet).

We visited the dam in 1998, the year after the terrorist attack centred on British tourists at the Luxor temple. No wonder there were very few people from Britain who visited Luxor temple that year. I wanted to go to Cairo from Luxor, but due to security reasons I was not allowed. Security was high and we had a long drive from Luxor to Lake Nasser, escorted by armed men.

The visit to Lake Nasser reminded me of my visits to the dam and reservoir on the Damodar River in eastern India. The Damodar used to be called the Sorrow of Bengal and, since the time of British India, several attempts have been made to control the floods using embankments. After Indian independence in 1948, the Damodar Valley Corporation (DVC) was set up and the dam, reservoirs and barrages were gradually built to control floods in Bihar (now Jharkhand) and West Bengal. The Damodar River originates at the Chota Nagpur plateau of the state of Jharkhand and flows about 368 miles (592 kilometres) through Jharkhand and the plain of West Bengal to reach the Hooghly River (the Ganges), ninety to one hundred miles south of Kolkata. The place is called Gadiara, and this is the meeting point of the Hooghly, Damodar and Rupnarayan Rivers. There is an old fort here, which was heavily damaged during a devastating flood in 1942. The fort is called Fort Mornington and was built by Lord Clive (1725–74).

In the early 1960s, my parents were invited to visit the Damodar Valley, and my two brothers, two sisters and I accompanied them. From Calcutta, we travelled by train: the Coal Field Express. This was a double-decker train and used to run from Calcutta to the coal belt region of eastern India. We got down at Asansol station, in the heart of the coal belt region. At the station, we met a tall Englishman, who was a friend of my father's but later on became close to me also. He was Mr

Miles Spaine, the general manager of a multinational company, and one of his factories was situated in the Asansol– Durgapur area. I was pleased to see the same tall man at Heathrow Airport when I arrived in England in January 1968.

Miles took me to various places in England and introduced me to various people, including his future Scottish father-in-law. He used to live in London and Chester. In 1968, I stayed on a couple of occasions with him in Chester. I remember that he drove from London in his car in the evening; we stopped overnight in Birmingham and then, the next day, at mid-morning, we started again and reached Chester by late afternoon. We used to talk about development, poverty and apartheid in South Africa, Moral Re-Armament (MRA) and others. MRA was the organisation whose movement was to make a new world by changing the individual and society based on four absolutes: honesty, purity, unselfishness and love.

I remember once he took me to Nottingham, which to me in my childhood was associated with Robin Hood and his legend. Naturally, I asked him, 'Where is Sherwood Forest?' At that time, Miles was not able to locate the place, as there was no Sherwood Forest. This was 1968. Most of the forest areas were long gone due to mining, industrialisation and urbanisation. The area that is left has now become 450 acres of Sherwood woodlands, and is developed as a tourist spot. I visited the place in 2003, and found that some of the finest and oldest oak trees of the country are located here; many are over a hundred years old. The most famous oak trees are the Major and the Parliament oak trees, which have a rich English history.

From Asansol station, we travelled twenty-four kilometres by car to the guesthouse where we were going to stay. Not long before, President Nasser had stayed there when he visited India. The visit to this dam was important to him as he was building the Aswan Dam at the time. The public relations officer of the DVC, who was a great friend of my father, told the story that President Nasser was a tall man and, when he stayed in the guesthouse, the original bed in the guesthouse was too small for him and so a special bed was made for him, replacing the original one.

My next brother Asoke took some photographs and produced a family album on this trip. Later on, he became a production manager in Indian Express Group of Newspapers and also a lecturer in printing journalism for Calcutta and Jadavpur Universities.

Two other rivers, the Koel and Sankh, rise from the Chota Nagpur plateau of Jharkhand. These two rivers interest me, as I worked in one of the most important steel cities in eastern India, a place called Rourkela, located at the confluence of the River Koel and River Sankh, surrounded by hills. It was a small village and became prominent after the establishment of the Rourkela Steel Plant in 1955. Rourkela is an industrial city now, with iron and steel plants, heavy machinery, fertilizers and chemical factories. Here, the Koel and Sankh meet together to form the River Brahmani, which flows 480 kilometres through the plains of Orissa and forms a big river delta with two other Orissan rivers, the Mahananda and the Baitarani, before flowing into the Bay of Bengal.

Besides Rourkela, industrialisation along the 480-kilometre river basin is responsible for the water and environmental pollution of the Brahmani River, both upstream and downstream. The river water is polluted as a result of discharging domestic, mining and industrial toxic effluents. In addition, sewerage water, floods in monsoon, and dry riverbeds in the summer are also responsible for the deterioration of water quality. All of this has had an adverse effect on human health and aquatic life. In 1979–80, while working in Rourkela, I discovered a cluster of a certain type of cancer especially in the Rourkela township, which might be linked to water or environmental pollution.

The monsoon brings extensive flooding to South Asia. July 2005 was one of the worst flooding disasters, caused by monsoon rain in Bombay (now Mumbai). The city is located on the west coast of India on the Arabian Sea. 37.2 inches of rain fell in a twenty four-hour period on 26 July, which superseded all previous records. The Western Ghats hill range, which lies thirty miles (fifty kilometres) from the coast, is responsible for high monsoon rain. Large areas of the metropolis of

Mumbai and the surrounding regions of the state of Maharashtra were flooded.

About 1,000 people died and the causes of death were drowning, suffocation (trapped in vehicles), landslides, stampede, electrocution and wall collapse. 25,000 sheep and goats and 2,500 buffaloes drowned.

The sewage system overflowed due to rain water and all water lines were contaminated. People were without electricity, power, tap water and waste disposal facilities. Twenty-five people died due to waterborne diseases and leptospirosis as a result of epidemic.

The transport system was most affected. Fifty-two local trains, 900 buses and 4,000 taxis were damaged. 37,000 auto rickshaws were ruined, and 10,000 trucks and tempos were grounded. Mumbai airport was closed for more than thirty hours and rail links were disrupted for ten days. Mumbai is the commercial capital of India, and the flood caused a stoppage of commercial trading and industrial activity for days. The flood caused direct losses of $100 million (Rs 450 cores).

Mumbai is famous for Bollywood and is the centre for the Hindi film industry. Bollywood produces 200 films annually and is the largest film-producing engine in the world. Flooding forced Bollywood production companies to cancel films. Bollywood is like Hollywood in America. Although the quality of the films is criticised, they have fans all over the world. Many young people in India, and also those of South Asian origin generally living in other parts of the world, have dreamed of becoming Bollywood superstars. Some of the Bollywood stars are so popular that they are identified with India more readily than statesmen or sports stars. Aishwarya Rai, Amitabh Bachchan and Shahrukh Khan are the famous Bollywood heroine and heroes who are displayed in London's Madame Tussaud's wax museum.

The Himalayan rivers are snow-fed and heavy rainfall increases the flow and floods to the Ganga–Brahmaputra basin. The Ganges, a holy river to Hindus, originates at the Gangotri Glacier, which is five miles by fifteen miles in size, is located in the Himalayas at a height of 7,756 metres and flows as the Bhagirathi River. At Devprayag in Uttarakhand, it joins with the River Alkananda to form the River Ganges. The Ganges leaves the Himalayas at Haridwar, travels over the plains of Uttar Pradesh,

Bihar, Jharkhand, West Bengal and Bangladesh for 2,510 kilometres (1,560 miles), and falls into the Bay of Bengal.

The upper Ganges runs from the Gangotri Glacier to Hardwar, emerging out of the ice cave and running through the valleys and rocks of the Himalayans Mountains with tumbling speed and strong currents. After Hardwar the Ganges leaves the Himalaya and falls into the vast plains of northern India in between Himalaya and Vindhya mountains ranges, travelling south and south- east. This is the middle Ganges that passes through cities like Kanpur, Allahabad, Varanasi, Patna, Monghyr, Bhagalpur and the Rajmahal hills. Near the Rajmahal hills, it changes its course southwards and enters into the West Bengal as a part of the lower Ganges. Near north-east of Jangipur, the first tributary of the Ganges, flows south as the River Bhagirathi and joins the Jalangi river at Nabadip to form Hooghly River, which ends at the Bay of Bengal after passing through Kolkata. The other tributary, which is the main branch of the Ganges enters Bangladesh as the river Padma until it is joined by the tributaries of the Brahmaputra, firstly the Jamuna and then the river Meghna which empties into the Bay of Bengal.

The Ganges is one of India's great rivers which I had always wanted to explore by boat or ship. Since the ancient times the Ganges and some of its tributaries were navigable and up until the nineteenth century regular steamer services used to run from Calcutta to Allahabad and beyond. However, with the construction of railways in mid-nineteenth Century the decline began of water transport on the Ganges. There are some services which are mostly located in Ganges delta on which I did travel on a number of occasions. However I could not find any means by which I could fulfil my dream of travelling from Kolkata to Allahabad and beyond by boat or ship until I joined the 'Pandav' cruise for its 1,280 kilometre maiden voyage of a fourteen days trip from Kolkata to Varanasi on 28 September 2009. This was the first time in two hundred years that a passenger vessel had made it all the way to Varanasi, although, the last sixty kilometres were by road because of technical difficulty and the pressure of time. Some of the passengers were also not well. Besides me, there were thirty- nine other passengers, nineteen from Australia, six from Britain, five from USA, four from Belgium, two passengers each from Canada and New Zealand and one person from Malaysia. While on board, 88 percent suffered from stomach infection such as diarrhoea. I also suffered from this as well as urinary tract infection. Later on I found out that I had the E.coli infection for which I was hospitalised.

E. coli is a common waterborne bacteria and it is a good indicator of faecal pollution and water contamination. It can cause gastroenteritis and urinary tract infection. The typical sources of infection include contaminated meat, vegetables, milk products, juice, sewage-contaminated water and bathing in contaminated water. The Ganges is well known for faecal contamination and some places the faecal count is too high even for the standard of bathing and no doubt the villagers or urban dweller along the Ganges may easily develop waterborne diseases if such water is used.

E Coli bacteria grow easily in the alkaline media. Throughout my Journey from Kolkata to Varanasi, I collected fifteen samples of Ganges water from fifteen different sites. Out of the samples, at least five samples showed very high pH (Alkaline) which supports that in some places Ganges is highly polluted with sewage.

Pollution in the Ganges is a big issue due to raw sewage, human and industry pollutants, as the river contents untreated sewage, debris, cremated remains and chemicals. The human pollutants are mostly from inadequate cremation as a result of unburned or partially burnt corpses and livestock. I was told that this is because of the lack of wood or inadequate burning material in some villages along the Ganges. There are certain cities and towns where there are crematoria and also more crematoria are being established, but the traditional cremation is still popular as many Hindus believe that it is sacred to be cremated on the Ganges river.

Industrial pollutions along the Ganges constitutes much smaller proportion of pollution in comparison to sewage pollution. Industrial pollution is mainly from the leather, textile, jute, paper, pharmaceutical industries, refineries, power plants and chemicals used in farming. Some of the recent reports on pollution suggest that there are risks of arsenic groundwater contamination in Ganges delta and middle Ganges plain. Other metal-related pollution also can not be ruled out as I saw many disused boats, ships or barges which had been marooned for some time.

To overcome the Ganges pollution, in 1985 the Indian Government launched the Ganges action plan, which was primarily focused on the treatment of waste and sewage. Since then India has erected a number of waste treatment plants, established more crematoria and spent US$330 million but the Ganges water is still not safe. It is muddy and some places remain dirty. It is good news therefore that in December 2009 the World Bank has agreed to lend India US$1 billion over the next five years to clean up the Ganges.

Hopefully, this money will be utilized effectively and efficiently.

Our cruising ship was built in Myanmar in the year 2004-2005. The ship is fifty -four metres long and twelve metres wide. The height is 12.50 metres and draught is 1.40 metres. Long distance upstream cruising in the Ganges is not easy. Our lower Ganges journey was from Kolkata to Farakka which is the Bhagirathi-Hooghly stretch of Ganges. We sailed through various curved river loops which makes the voyage treacherous. Some sailor call it the 'snaky way'. The shifting channel, sudden appearance and disappearance of the chars (sandbars) make it difficult to navigate the Ganges. In the past the land and palaces of various kingdoms were lost in the river bed. In some places the Ganges is not very wide; only 1,200 feet and it is also occasionally quite shallow, with a depth of less than two metres. Our ship did not have any instruments for measuring the depth of the water. Most of time, our ship was guided by the pilot boat, which belonged to the Inland Waterways Authority of India. The pilot boat had a water depth monitoring system and the staffs were well-trained. One of them showed me on the computer that how river bed had changed within three weeks, as we were not able to use the same channel route because of shallow water. In spite of this, a couple of times we were stuck on shallow water in the middle Ganges. It seems that the Ganges channel changes its course very quickly. Of course it is also largely depends on the time of the year when one is travelling.

Our ship also had to negotiate low-hanging power lines, cable lines, low bridges, and high/low tide. We entered the middle Ganges near the Farakka Barrage. In order to do so we had to pass through the Farakka feeder channel and entered the middle Ganges through a lock gate. In between the two gates of the lock, I was surprised to see plenty of water hyacinths. This lock gate is not often in use, usually opening two to three times a month for cargo vessels carrying cement, fertilizer, crude and audible oil to pass. No wander that the water hyacinths, which usually grow in the stagnant water, were thriving. In Ganges delta it grows widely and sometimes is used as an adaptive measure against rising flood levels.

The middle Ganges is quite wide: in some places it is more then 10 miles and when the Ganges floods it becomes fifteen to twenty miles. In the middle Ganges near the Rajmahal, we suddenly found that the ship was running with one engine instead of two, as one of the propellers had been damaged by submerged debris.

The speed of the ship was initially eight kilometres per hour and this was reduced to nearly half of its capacity until a tug boat was attached. The speed picked up with the support of tug boat, though not fully. We also had slight scary experience in the middle Ganges before Patna, when the crew had difficult time to navigating in the dark night due to storm, wind and cloud. Moreover, we had to rely on one anchor as the chain of the other anchor suddenly snapped.

Despite all those pitfalls, I enjoyed the cruising as the scenery along the Ganges was beautiful. I saw the beauty of the Ganges with its changing colours in scorching sun, in dawn, in dusk, in rain, in storm, in breeze and in cloud. On the char (Sandbanks) and on the river banks, the *kashphool (*Bulrush) plants with their white flower were dancing and waving in the autumn breeze and the sun rays were peeping through. The rows of chimneys of the red brick kiln looked like imarets flashing in the setting sun. The views were spectacular. Along the banks, we saw the fertile soil of the Ganges where the crops such as rice, jute, wheat, burley, sugarcane were growing. In some places the cultivation had been modernised: I saw some farmers using tractors, spraying chemicals and irrigating the land by pumping water from the Ganges. Cattle, horses and water-buffalo were gazing on the fields. Men, women and children were bathing. Water-buffalos were being washed. Along the Ghats and the banks we saw fairs, festivals and ceremonies were taking place as well as the cremation of the dead. Religious prayers were ongoing. Men, women, *sadhus* and priests were chanting or singing. We saw villagers standing on the river bank, waving, friendly and welcoming when we stopped for visits. I visited Kolkata, Chandanagore, Kalna, Murshidabad, Malda, Rajmahal, Sahebganj, Munger, Vikramshilla University, Patna, Ghazipur, Sarnath and Varanasi. In most places we were surrounded by reporters, journalists, television crews; some of us had to give press or television interviews as we were the first passengers on this maiden voyage.

Along the Ganges we saw some of the rare birds and mammals that inhabit it such as river dolphins, dears and antelopes. Fishermen, fishing nets and fishing boats were common sight on the lower Ganges or Ganges delta but less so in the middle Ganges. There we saw the river banks or the char with beautiful white sands, glittering in the sun so it looked as if we were by the sea, sailing alongside the virgin sandy beeches.

At the mouth of the Ganges is the biggest delta in the world, wherein lies the Sunderbans, 2,585 square kilometres of mangrove swamp, with a wildlife sanctuary extending into West Bengal and Bangladesh. Part of Calcutta was once within the Sunderbans area. This vast land mass was once full of swamp and forest. The name 'Sunderbans' comes from the trees, sunderi, that dominate the land mass. This is where Royal Bengal tigers live and roam freely from island to island. In the seventeenth century, tigers used to be found where the Kolkata Park Street Cemetery is now located. With years of urbanisation and deforestation, the mangrove swamp and forest was ruined; the remaining area is now the Sunderbans Wildlife Sanctuary, which is a protected area and a World Heritage Site. Although debatable, the tiger population numbers about 280 on the Indian (West Bengal) side and about 500 on the Bangladeshi side, based on the 2004 census. These are the only man-eating tigers found in the world that swim and drink saline water. Each year, at least twenty people are killed by them. The tigers also survive by eating human and animal corpses that are found in channels, rivers and estuaries as a result of natural disasters like floods, tropical storms and tidal waves, which are not uncommon in this delta region.

Woodcutters, honey collectors and fishermen, in respect of their own religion, worship the common goddess of forest, the *Bonbibi*, before entering into the forest on their way to their hazardous occupations.

Royal Bengal tigers attack from the rear, and so nowadays villagers wear masks painted with human faces on the back of their heads; a tiger is less likely to attack the worker, as it thinks the individual is watching it. I visited the Sunderbans area in early 1960s, but not the deep forest regions. At that time, it was not a sanctuary. More recently, on the 3 and 4 January 2008, I visited the West Bengal side of the region again – it is now a tiger project area. We sailed round the Sunderbans delta on a local ship and stayed two days and one night on board. Although I was disappointed that I was not able to see a single tiger, I was thoroughly impressed by the beauty of the true mangrove forest and was able to identify some of the mangrove plants where tigers often hide or roam about.

To get into the tiger project area, one needs permission and also has to have the appropriate agency along with an armed escort as guide. The West Bengal tourist board also runs the tour and my visit in January 2008 was through the tourist board, without any armed escort. Most of the journey is usually by boat, launch or ship.

Glaciers in most areas of the world are receding, and so are the glaciers of the Himalayas. The Gangotri Glacier has already receded, as indicated by satellite and historic data. This will have an impact on water supplies, flooding, agriculture, hydroelectric power, water transportation, coastlines and the ecological habitats of the Ganges Valley.

The Ganges Valley is thickly populated and there is a joke that every time it floods, there are more corpses and more skeletons to export. It is true that, to learn anatomy, natural human skeletons are the most important aid for medical students, and for this purpose – at one time – human skeletons were exported to Britain from India. Nowadays, the export of skeletons from India is completely banned, and I heard from an anatomy professor in a medical college in India that they are also finding it difficult to find human bodies for dissection. Whatever the joke is, India is trying to control the floods by building dams.

On the Ganges, there are two dams; one in Hardwara in Uttarakhand and other one is the Farakka Barrage located in West Bengal, where the Ganges is divided into Padma and Hooghly. I visited both Hardwara and Farakka. There are two more dams coming up on the northern Gangetic plain, which are in the planning stage.

India has 9 per cent of the world's dams. Pundit Jawaharlal Nehru (1889–1964), who was the first prime minister of independent India, gave priority and importance to building dams, hydroelectric power and irrigation projects. So the dams were built on rivers: Damodar in Bihar (now in Jharkand), Hirakhud on the Mahanadi in Orissa and Bhakra Nagel in Punjab. It is said that the Hirakhud Dam is the first justifiable dam in India for flood control, which I visited at both pre-construction and post-construction stages. It is now controversial because critics say that floods were more frequent after the Hirakhud Dam was built. Sometimes, an emergency release of water is responsible for a downstream flood. Of course, the regular reservoirs' sedimentation prevention and high dam construction will no doubt prevent further flooding. Similar problems arise from forced discharges of water from the Bhakra Nagel Dam and Damodar Valley.

In the meantime, despite some political and religious outcry, the controversial large dam project on the River Narmada, Madhya Pradesh, is ongoing.

Since 1953, India has spent many billions of dollars on flood control by building dams, embankments and river canalisation, but still, from time to time, one can see or read news like: 'More than 800 people have been killed and millions displaced by floods and landslides in India, Nepal, Bangladesh and Pakistan.' The water of the major rivers that flow from the Himalayas is a great issue in India, Nepal, Bangladesh and Pakistan, especially in terms of water-sharing and irrigation.

Abnormal monsoon rains in July and August 2010 were responsible for catastrophic floods in Pakistan and India. Torrential rainfall of more than 200 millimetres (7.9 inches) triggered the flood causing maximum damage to the province of Khyber Pakhtunkhawa (formerly the North West Province), Punjab and Sindh in Pakistan. The flood water flowed into the major rivers and its tributaries, canals and nullahs.

The main river affected was the Indus, which is 3,200 kilometres (1988 miles) long, and originates in the Tibetan plateau, near Lake Manasarowar. It runs through both Indian and Pakistan-administered Kashmir and then flows through the plains of Pakistan from the north to the south of the country before it enters the Arabian Sea near Karachi.

The Indus is largely fed by the snows and glacier of Karakorum, the Hindu Kush and the Himalayas. In winter its flow is diminished and during the monsoons (July to September), it floods the plains. This year (2010) extremely heavy monsoon rains triggered the floods, which resulted in the swollen Indus river bursting its banks and breaking the current flood controlling system that is, the river defence embankments and levees. This destroyed many towns, villages and cultivated lands. According to one figure, 1,802 people died, 2,994 were injured and 1,894,530 homes were destroyed. 1,000 bridges, 5,000 miles of road and rail were washed away. Almost 21 million people were affected and more than 100,000 square kilometres were invaded by the flood waters of which 20,000 square kilometres were agriculture lands. There was a question as to how the situation was made worse by diverting the flood waters from certain properties and destroying the neighbourhood.

According to Pakistan officials, this flood ruined crops worth 281.6 billion rupees ($327 billion), destroying rice, cotton and sugar. Many people lost livestock and their livelihoods. Many were vulnerable to malnutrition and other health problems. Respiratory and diarrhoeal diseases have been reported by WHO and there was the threat of waterborne diseases, including cholera. Initially the speed of the disaster response was poor and this undermined the gravity and seriousness of the flood.

The floods in Leh, India occurred on the 6th August 2010. Leh city is the largest town in the Ladakh region and is on the ancient Silk Route from China. Nestled in the mighty Himalayas it is situated at the height of 3,500 meters above sea level, in the extreme north of the country. Large parts of Leh were devastated when flash floods were triggered by a cloud burst. Gushing water and mud killed at least 193 people. Thousands of people were injured, missing and homeless. Many buildings, roads and bridges were washed away or damaged and the local bus station was flattened. Communication systems were badly affected and the local airport damaged. Many tourists including more than 3000 foreign visitors were stranded. The rescue operations were carried out by the army. Emergency camps had to be set where relief supplies could not reach in time because the main highways into the region were cut off by landslides. Rumella and Rahul went to Leh to report for the BBC, one week after the incident and sent me some photographs; one of which is at the end of this chapter showing the devastation after the flood waters had subsided. Most of the damage was in old Leh and the surrounding villages of Choglamsar and Shapoo. In Choglamsar, more then 40 homes swept away by the overflowing river Indus. Sand bags were used by the Indian Army to stem the flow of the river preventing it from causing further damage.

Bangladesh is well used to flooding. Besides coastal flooding, monsoon flooding usually hits Bangladesh. Bangladesh lies on the Ganges–Brahmaputra river delta, which is the biggest delta in the world. Some parts of the delta are so vast that it looks like the sea. To cross the delta by paddle steamer boat, and then by country boat, to the then-undivided Bengal (or just after Bengal's partition), was a great childhood experience for me. My childhood memory in the mid- or late 1940s is of wood-panelled cabins and a deck full of passengers. People with their luggage were hurriedly getting on and off at their ports of call, as the crew shouted their destinations. Ferries, guided by their sareng (navigators), were passing by and I could also hear the sound of warning sirens. I could see rafts made of bamboo that were carrying buses, cars and bicycles across the river. The rising sun, which looked like a red ball, was on the autumn horizon. We passed by fishermen and their fishing nets, and saw boatmen tugging their boats against the low tide, and the long chains of floating timbers, guided by the country boat. On the river banks, we could see paddy fields and the golden paddy dancing on the gentle breeze, as well as bamboo groves and pineapple groves. On one side, the riverbank was vanishing in the water and on the other side, new land was emerging.

Good harvests and bumper crops flourish after every major flood in Bangladesh.

The year 2004 saw the worst river flood in Bangladesh in recent memory. Nearly three quarters of the country was under water and twenty million people were stranded. Embankment failure was very common. To prevent flooding, canalisation of the heavily sediment- laden river is needed. This is not easy as, in the low-lying delta, migrating rivers and sudden river channel shifts occur naturally.

To prevent flooding in the Indian subcontinent, step dams that could be built along the Himalayas have also been suggested. India is thinking of connecting thirty-two rivers with channels to prevent chronic flooding and drought. There is also the question of reviving dead or nearly dead rivers and channels. All are ambitious proposals and are not easy to implement. However, to prevent flooding and its damage in the Indian subcontinent, more co-operation is needed from India, Tibet, Nepal, Pakistan and Bangladesh. Tibet's or China's involvement is necessary, as some of the great rivers of the Indian subcontinent originate from the Tibetan plateau.

The USA has spent over $25 billion on 500 dams and 16,000 kilometres of embankments and river canalisation in the war against floods, but the USA is still not able to overcome damage to property and the number of dead due to flooding that occurs annually. The Mississippi River, which is known as the 'Father of Waters' by Native Americans, is said to be the queen of floods as unexpected floods occur and cause catastrophic damage.

Another of the USA's great rivers is the Colorado River, which flows from the Rocky Mountains to the Gulf of California. Its 1,450-mile course passes through deserts, canyons and mountains. The most famous is the Grand Canyon, which I visited in spring 2006.

The Grand Canyon of Colorado is made up of many canyons and gorges. It is 277 miles long, eighteen miles wide and one mile deep where the Colorado River flows. It took twelve million years for the Colorado River to carve out the Grand Canyon, making a magnificent landscape in the earth. It is no wonder that it is one of the seven natural wonders of the world.

Every spring, the Colorado River floods along its route, especially in low-lying areas. Trying to control the flood, dams have been built on the Colorado River, of which the most famous is the Hoover Dam, which I saw on my visit to the Grand Canyon.

It is a concrete gravity arch dam on the Colorado River, surrounded by rocky canyon walls, on the borders of Nevada and Arizona. It took five years to build the dam, and it was completed in 1936. It is the main source of electricity, irrigation and flood control for south-west America.

From Las Vegas, I took a flight to Grand Canyon West Airport by a small plane, a forty-five-minute flight. The flight was smooth and the weather was very good. It was a clear day, the aircraft was flying low and the Grand Canyon was clearly visible. As we flew over the Grand Canyon, I saw thousands of canyons with a variety of shapes and colours, thousands of gorges, the Colorado River and Hoover Dam, desert and Table Mountain. After landing at Grand Canyon West Airport, I had a helicopter ride to the floor of the Grand Canyon, where I boarded a small boat to cruise the Colorado River, and then reboarded the helicopter and ascended the canyon wall. While descending and ascending 4,000 feet, I had a spectacularly close view of the canyon rocks and walls. While sailing through the gorges, we saw the splendour of the rocks and saw how the river carved the canyon. The river water was muddy and the old, famous statement about the Colorado River – 'Too thick to drink and too thin to plough' – probably still applies.

On the canyon wall, in some places, I saw some graffiti, which was Native American art, and this was pointed out by our boatman. I also took a bus trip along the rim of the canyon and stopped at two places, at the north and south rims, where we saw spectacular views of the Hualapai Rim plateau. I could see the Colorado River carving through the red, white, buff, grey, yellow, orange, brown, pink and black rocks of the canyon; it sometimes changed its colour with the sun and clouds.

I also visited a Native American village and saw their dwellings, dresses, jewellery and a show.

Certainly, I saw splendid views of the grandeur and colour of the Grand Canyon and the Colorado River. At the end, as usual, I lost my fleece, which is not uncommon for me, as quite often I lose something on my travels or on holiday.

Widespread flooding causing catastrophic damage in Central and Southern America is also not uncommon. The Amazon River in South America is the second longest river in the world, after the Nile in Africa. It took a long time to establish the source of the Amazon and, in 2001, it was confirmed that it originates as a stream from a peak in the Peruvian Andes. The peak is called Nevado Mismi, and the height of the peak is 18,363 feet (5,597 metres). From its source in Peru, the river flows about 6,400 kilometres (4,000 miles) through Brazil before entering the Atlantic. The widest point of the Amazon is eleven kilometres (6.8 miles), and when the Amazon basin floods, it can spread up to forty kilometres (24.8 miles). The Amazon is subject to severe flooding, and seasonal rains are responsible for this. To control the floods, there are eleven dams, although I gather there are more coming up.

The Amazon basin is famous for the Amazonian rainforest, which begins east of the Andes and covers most of the Amazon basin. It is the largest rainforest in the world and its extremely wet climate helps its natural ecological system. In a time of global warming, the survival or conservation of the Amazon rainforest is important, as it has an enormous capacity to absorb carbon dioxide.

After the Amazon, the Parana River is the next longest river in South America, originating in southern Brazil at the confluence of the Paranaiba and Grande Rivers. It then travels 2,485 miles through Brazil, Paraguay, Argentina and Uruguay. On the way, it meets with the Iguazu and Paraguay Rivers and merges with the River Uruguay to form the Rio de la Plata, before entering the Atlantic Ocean. In its long journey, it encounters waterfalls, dams and a delta. The Iguazu Falls is the most famous falls in South America, and we saw spectacular views of it from both the Brazilian and Argentine sides when we visited it in August 2007.

The waterfall system is made of 270 falls along 2.7 kilometres (1.67 miles) of the Iguazu River. The individual falls vary from sixty- four metres (210 feet) to eighty-two metres (269 feet) in height. Several walkways and footbridges take one very close to the falls, so that one can hear the noise, feel the power and see the volume of the water. Several floods have destroyed the footbridges, but they have been rebuilt every time. This was the first time I had heard about falls being flooded, and some say that this is due to the construction of the Itapúa Dam. There have been cycles of droughts and floods. There were floods in 1992, 1997, 2005 and May 2007. Along the course of the river, near the confluence with the Paraguay River, there is another dam called Yaciretá. Both Itapúa and Yaciretá Dams produce hydroelectric power, which I understand contributes towards Paraguay's status as one of the world's largest hydroelectricity exporters. The Parana River, along with its tributaries, creates a massive delta of 14,000 square kilometres. The length of the delta is 320 kilometres, and its width varies between eighteen and sixty kilometres, making it a long flood plain. We explored this area on a guided boat trip in July 2007.

Russia is great at building high, strong dams and large reservoirs. The Nurek Dam on the Vakhsh River, Tajikistan, is the world's tallest dam, with a height of 300 metres (984 feet). The strongest dam is the Sayano-Shushenskaya Dam on the Yenisei River, designed to bear a load of eighteen million tonnes. The largest man-made reservoir is on the Angara River in Russia, called the Bratskoye Reservoir, and the total area is 5,470 square kilometres (2,112 square miles).

However, the largest dam in the world is in the USA, which is on Ten Mile Wash in Arizona, and it has a capacity of 209.5 million cubic metres (274.5 million cubic yards).

China has 45 per cent of the world's dams and is still fighting to control flooding. Huanghe or Hwang Ho (the Yellow River) is called 'China's sorrow'. This is the muddiest river in the world. It flows from Tibet to the Yellow Sea and bursts its banks often, resulting in extensive flooding that causes millions of deaths, homelessness and destruction. In the last fifty years, China has built hundreds of dams along the Yellow River, whose 3,400 miles course through nine provinces. River water has been diverted for the large-scale irrigation of China's interior provinces, and I gather that China is planning ten more dams on this river. It seems that the Chinese are making a great effort to control the great river by building dams. However, there are environmentalist worries that the Yellow River will soon dry up due to over-extraction of water, the discharging of toxic materials from cities and factories and severe drought.

Mentioning the over-extraction of water reminds me of the Aral Sea, located on the Uzbekistan–Kazakhstan border (previously part of the USSR). Since 1960, the Aral Sea has decreased in size from 68,000 square kilometres (26,300 square miles) to 26,800 square kilometres (11,000 square miles) and 80 per cent of the volume of water in the lake has also disappeared. This is a result of water being diverted from the two rivers (the Amu Darya and the Syr Darya) that feed the lake. The water is used to increase the agricultural use of the lands around the Aral Sea. This has not only caused the sea to sink but has also increased its salinity, resulting in the end of commercial fishing in the waters in 1979.

Another great Chinese river is the Yangtze River, which we cruised up during our visit to China in spring 2004. This is the longest river in Asia and the third-longest river in the world. The river is 5,470 kilometres (3,400 miles) long, originating from the Kunlun Mountains in the south-western region of the Qinghai Province of China (Tibet plateau). It flows south and then northeast and east across central China to its mouth, located in the East China Sea, north of Shanghai. The river has over 700 tributaries. We cruised along not only the Yangtze River but also some of its tributaries and canals at Nanjing, Wuxi, Suzhou and Shanghai.

In China, I was with my wife. With us from Britain, there was a group of people which included Dipak and his wife, Jaba, and Jyoti and his wife, Pampa. Dipak Datta, whom I knew very well from my college days, was working at that time in Sheffield as a consultant in rehabilitation medicine, and Jyoti Datta worked in Billingham, Cleveland, as a senior general practitioner.

Along the river banks, we saw rice and wheat fields and, on the river, we saw Chinese fishing nets. The fishermen were catching fish and in some places the river was congested with barges and country boats. I gather that, in this part of China, people are very much dependent upon river transport. Fish from the river is a local delicacy. The Chinese eat many things that people in other countries do not eat or enjoy, and this I saw when we visited a local market. In restaurants, visitors can choose fish or other marine products, which can be cooked in front of the customer. Some of us did not like the food in China, as the taste was different from Chinese food that we eat in Britain, the USA or India. Pampa was vegetarian and she did not enjoy the Chinese vegetarian food as most days she had the same types of food. She was the only vegetarian in the group but she ate fish. The restaurant people could not understand how the vegetarian became a fish-eater, and there was a problem of sharing food as the food was usually served on the table for everyone to share. Pampa said, 'I am fed up, I am hardly getting a proper meal'. On our China trip, lodging and board were included. No wonder she was cross with the travel agency, although our accommodation was superb, and this in spite of me being a disabled traveller.

Unfortunately, I had to travel China with a walking stick, as my right knee cartilage was damaged due to an accident and I was waiting for an operation. I had restricted mobility, but it did not put me off visiting China. The accident happened at our home in England. I was coming down the stairs carrying my grandson Rishav, when my right leg gave way, resulting in a knee injury. This no doubt had a tremendous impact on my climbing and walking on the Great Wall, the famous architectural landmark and visitors' attraction. The Great Wall, which protected ancient China over the years, was first built during the period of 770–476 BC and building continued on and off until the seventeenth century AD; a major contribution was during the Qin dynasty (207–201 BC). The wall is probably between 1,500 miles (2,400 kilometres) and 4,000 miles (6,400 kilometres) long, and is twelve to thirty feet wide. Originally, it was built with wood, stone and compacted mud, but later on brick was used. It was no doubt a strong structure. However, I had to restrict myself to walking only a short distance on the top of the wall, although I enjoyed the spectacular view of the long chain of the Great Wall.

Like the Yellow River, the Yangtze River is prone to catastrophic flooding, drowning China's best farmlands and leaving a number of people dead, homeless and subject to starvation. In 1954, when my father visited China, I heard there was a flooding of the Yangtze River that killed three million people. In spring 2004, when we visited China, there was no flood, but six years ago, in 1998, not far from where we cruised, there was a flood on the Yangtze River, and 150,000 Chinese soldiers were sent in to sandbag the river and the dykes. It was said that this was the worst flood in living memory, and it claimed 3,000 lives, wiped out entire villages, isolated cities and destroyed million of hectares of crops. Deforestation is blamed for this devastating flooding.

To prevent this type of calamity, the Chinese are building the massive Three Gorges Dam on the Yangtze River, which will probably solve some of the long-standing Chinese flood-related problems. This will also generate hydroelectricity, and China is expecting that the electricity generated from it will provide one ninth of China's total power supply. The construction started December1994 and expected for full operation in 2011, though the main dam was completed in four years ago. The

Chinese authority was happy as China's controversial Three Gorges Dam withstood the high flood water of Yangtze River in July 2010.

Chinese literature, art, music, food, culture and civilisation emerged in the valleys of these two rivers and, as such, I was proud to visit one of the oldest civilisations in the world, as the country is opening up to foreign visitors.

When we visited Europe, Salil and his wife Marianne used to take us to the River Rhine.

Salil Chandra is my wife's cousin who used to live in Germany. I love river-cruising in Europe and I cannot remember how many times we have cruised on the Rhine and the other rivers of Europe. The mighty Rhine originates in the Alps and flows nearly 900 miles from Switzerland, through the Black Forest, across France, Germany and Holland, ultimately falling into the North Sea. I have seen the beauty of this river at various places and have also enjoyed the romantic scenery of the middle Rhine. Sometimes, one notices the red flood signal or flood warning while cruising. There are three types of warning. 'Red' signifies severe flood warning. 'Brown' signifies flood warning and indicates that flooding of homes and businesses is expected. Lastly, 'yellow' signifies flood watch and indicates that flooding of low-lying land and roads is expected.

Europe has many rivers and the most famous are the Rhine and Moselle, the Danube, the Elbe and Vltava Rivers, the Seine, the Rhone, the Yonne, the Douro, and the Po.

River-cruising in Europe is great and the most remarkable one was a cruise from Amsterdam to the Black sea which we undertook in July/August 2008, lasting twenty-two days. Our ship was Ms Johann Strauss. We sailed across Europe, cruising along the Rhine, the Main and the Danube, visiting Germany, Austria, Slovakia, Hungary, Serbia, Bulgaria and Romania. This was possible because the Main –Danube canal has now been opened for navigation, linking the North Sea with the Black sea. The task of connecting the Rhine with Danube was begun in the middle ages, but the idea of Charlemagne and Ludwig the First was not fully materialised until the year 1992 when the canal between Bamberg and Kelheim was completed.

In this 3391 km voyage we crossed seventy locks, two canals, three rivers and passed under the 161 bridges and visited the wetlands of the Danube delta. We boarded the ship in Amsterdam and sailed the Amsterdam-Rhine canal for fifty kilometres before reaching the Rhine. At Mainz, we left the River Rhine and entered the River Main. We cruised on the River Main until we reached Bamberg where we entered the Main-Danube Canal. We cruised 166 kilometres on the Main-Danube canal and entered the River Danube after Kelheim.

Here the River Danube is 2,411 kilometres from the Black sea. Our ship cruised up to Tulcea which is the gateway of Danube delta, only 71 kilometres from the mouth of the delta. As soon as the delta began, the landscape along the river was changed. We saw some rare birds like stroke, cranes, and kingfishers. We also travelled to the Black sea port of Constanta from Cemavoda where the sixty-four kilometre-long Danube canal starts, connecting the Danube with Constanta, Romania's biggest port.

In our journey from Amsterdam to Tulcea, we passed through seventy locks: two in Amsterdam- Rhine canal, thirty-four in River Main, sixteen in Main – Danube canal, and twenty-five on the River Danube. From the low-lying Netherlands, we reached the highest point on the European waterways in the Main-Danube canal at the height of 1,332 feet above the sea level after rising with the help of forty-seven locks. Then, by the aid of twenty-three locks, our ship gradually came down to the level of 285 feet, a drop of 1047.78 feet. These locks are mostly navigation locks, but some of them have roles as flood locks as well. These are more important as flooding in the River Danube is quite common and this was quite visible as we saw the flood level marking of the Danube when we visited some of the cities and towns along the banks of Germany, Austria, Hungary, Romania and Bulgaria. The flooding of the Danube in April 2006 reached its highest level in some parts of Romania and Bulgaria in one hundred years as a result of heavy rainfall and melting snow.

While cruising in Europe we passed through a gastronomic landscape, lush vineyards, beautiful riverside towns and cities, castles and fortresses. Over thousands of years, these rivers have witnessed the fall and rise of emperors, the First and Second World Wars, and have contributed to European civilisation and renaissance, the development of trade, commerce, art and science. It is not unusual from time to time for torrential rains and flooding rivers to be responsible for overwhelming the river dams, destroying houses, historical buildings, cities and art.

The low-lying Netherlands has always had the threat of flood from the North Sea, as well as from swollen rivers like the Rhine and the Meuse. Over the centuries, dykes and drainage systems have contributed to the prevention of the catastrophic disaster of the submersion of the Netherlands under water or sea. Sailing through the drainage system leading to the sea was a great experience for me, and I have done it a couple of times.

The last catastrophic flood in Europe was in summer 2002, when flooding hit Germany, Austria, Hungary, the Czech Republic, Slovakia and Romania. The most affected cities were Dresden and Prague. Thousands of people were evacuated and historical buildings and museums were damaged.

Prague, the capital of the Czech Republic, is one of the finest cities in Europe, and stands on either side on the Vltava River bend. The Vltava River is the longest river in the Czech Republic and rises from the Sumava Hills near its southern border with Austria.

The Vltava River, along with its tributary, Berounka, drains most of southern and western Bohemia. From ancient times, Prague has been prone to frequent flooding due to this huge drainage basin. The Charles Bridge is the favourite tourist spot over the Vltava River and the bridge was constructed in 1357 to replace the Judith Bridge, which was washed away by floods in 1342. Severe flooding also occurred in 1784, 1827, 1845 and 1890. In the 1950s, the Slapy Dam was built twenty kilometres south of the city to control the Vltava River.

However, flooding in the Vltava River is still not controlled, as indicated by the floodwaters which hit the city in August 2002. Heavy rainfall caused flooding from thunderstorms in southern Bohemia and it seems that about 9.7 cubic kilometres of rain fell over nine days. The riverbanks burst and water reached five metres in height, flooding the eastern and western banks of the river. Nineteen people were killed and eighteen metro stations were damaged; the metro was closed for more than six months. Prague is proud of its arts and architecture, which goes back 1,000 years and survived the Second World War. We saw Romanesque, Gothic and European Renaissance-style buildings, sculptures in the streets and squares, mosaics and painted decoration on the buildings and numerous art galleries. Some of these were damaged during the August 2002 floods. The damage caused by this catastrophic flood cost 2.4 billion Euros.

When my wife and I, Salil and Marianne from Germany, visited Prague two and half years later, in 2005, we took a walking tour in Hradcany and our guide pointed out the marks on the wall of a house at Malostranska that reached five metres in height during the flood in August 2002.

Most of the historical buildings, museums, arts and treasures were restored, and tourists are once again able to enjoy the art and architecture of this beautiful city.

However, I had never seen a chapel decorated with human bones or skeletons until we visited the Ossuary of the Sedlec Monastery, which is near to present day Kutna Hora, sixty kilometres away from Prague. Those bones that we saw were from abolished graves. They were not from flood victims, but were the victims of the great plague in AD 1318. 30,000 bodies were buried at the time of this plague.

It seems that when Europe floods, it loses many art treasures, but when floods hit Asia, it kills people. This might be due to the greater magnitude of floods, larger populations and greater presence of poverty in Asia.

Regardless of such outcomes, on the subject of flood prevention and protection, a greater emphasis has to be placed on flood risk management rather than disaster response and management.

Public awareness, education, training, appropriate weather forecasts and the role of the media are important aspects for prevention of floods and should not be ignored. Science, research, technological developments including flood protection products are crucial. Climate change related to flood risks also needs to be explored. Finally, flood-related health risks in the form of direct and indirect, immediate and delayed, physical and mental issues need to be addressed so that appropriate strategies can be undertaken to reduce the risks.

The main flood-related physical health risks are drowning, electrocution, injuries (minor to fatal), skin disorders (chemical related) and contaminated water-related gastro-intestinal disorders.

Flash flood disaster in Leh, India-2010

Chapter Five

Storms, Tidal Waves and Floods

Hurricane

Most people in the world have experienced a storm or wind in a mild form, but not the big or high winds. When a mild storm or winds become strong or high then they are called hurricanes, typhoons, cyclones or tornadoes. The name varies according to its severity and its geographical location. They are formed when very high winds circle around an area of low atmospheric pressure. So, hurricanes, typhoons, cyclones and tornadoes are nothing but the manifestations of a storm with a strong wind.

Coastal areas suffer from cyclones, hurricanes and typhoons, whereas inland areas suffer more from tornadoes.

Hurricanes occur or originate in the Atlantic Ocean or Atlantic basin, and are named after the storm god of the Caribbean Indians, Hurricane.

The strength and speed of a hurricane is measured by the rate of wind per hour, and the intensity is rated on the S-S (Saffir/Simpson) category (Table 2), the scale ranging from one (weak) to five (most intense).

Table 2: Saffir/Simpson (S-S) Scale

Category	**Wind Rate**	
	Km/h	**Mp/h**
One	119–153	74–95
Two	154–177	96–110
Three	178–209	110–130
Four	210–249	131–155
Five	>250	>155

The name of a hurricane is usually chosen from lists selected by the World Meteorological Organisation, which alternates the name by male and female gender.

Hurricane Gilbert in the Atlantic, which swept the Caribbean and American south in September 1989 with a wind rate of 350 km/h (218 mph), was the strongest hurricane of Category Five that has ever been recorded. Hurricane Andrew in the Atlantic swept the Bahamas, southern Florida and Louisiana in August 1992 with a wind rate of 257 km/h (160 mph). It used to be the USA's most expensive natural disaster and the estimated cost of the damage was up to $20 billion.

In 2004, four hurricanes hit the Florida coast in six weeks and they were Frances, Charley, Ivan and Jeanne. This was the first time since

1984 that the state of Florida received quadruple strikes in a single hurricane season (which runs from 1 June to 30 November in each year), although Hurricane Charley hit in August and the other three in September 2004. All originated in the Atlantic and created havoc in the Caribbean islands, leaving a trail of devastation that killed people, destroyed homes and battered the coastline. There was severe flooding, fallen power lines, trees and debris, and roads cut off, resulting in a great economic effect on the tourism industry. Some islanders survived by taking shelter in the caves, and looting was not uncommon. Those four hurricanes battered Grenada, Jamaica, Cuba, Haiti, the Dominican Republic, Panama, the Cayman Islands and the Bahamas. Hurricane Ivan was 155 mph, which is Category Five; Hurricane Charley was 145 mph which is Category Four Hurricane Jeanne was 115 mph, which is Category Three, and Hurricane Frances was 124 mph, that is Category Two. All the hurricanes started with a deadly storm, but lost their strength in their pathway, so that, when they reached the American coast, they were not that powerful. For example, Hurricane Jeanne was Category Three, but became Category One when it reached Florida, before leaving more than 2,000 people dead in the Caribbean. Many people became homeless, and residents who had left their homes, on return, found that some of their homes were destroyed. Injuries, fatalities were not uncommon.

Hurricane Ivan was the strongest hurricane of the 2004 Atlantic hurricane season. It killed 121 people, of which fifty-four were from the USA. Hurricane Ivan was also responsible for $14.6 billion of property damage in the USA alone. Hurricane Ivan, which was Category Five, became Category Three when it reached the USA coastline. Creating a huge wave of water, Ivan made landfall around 2 a.m. on 16 September on the US mainland, near the shores of the Gulf of Mexico. Most damage was observed in Pensacola, Florida, due to winds, rain and waves.

I heard from one of the families who live in Navarre, twenty miles from Pensacola, where the eye of Hurricane Ivan passed through. Sourov Mascharak and his wife Haimanti, with their two children, Rohi, aged fifteen, and Noami, aged eight, were the only family within a fifteen-mile radius who stayed on and experienced the terrible night of 16 September. Sourov is an ex- US Air Force serviceman and now has a building construction business.

Most of the local population were old people and military personnel. On the night when Hurricane Ivan struck the Navarre, the Mascharak family were at home and they had to take shelter in the attic to avoid a worse outcome. They went up at around 10 to 11 p.m. They carried nothing but water, a torch and fruit (banana). They stayed there until 5 a.m. Throughout the night, they heard the sounds of shattered roofs from gusts and flying projectiles. Wind, rain and waves caused havoc. They could see from the attic with their torch how the sea and rainwater was getting into the house, the water level reaching six feet, and then eight feet. They saw two waves flooding the premises, and a third wave took away most of their home utilities. They could see snakes and other creatures in the floodwater down below. The water was black and contaminated. They were extremely scared, especially his wife and their younger daughter. Haimanti started to talk unrealistically and non-rationally. However, Sourov and Rohi kept their stamina alive and gave moral support to the others.

When Ivan's destructive power was over, they came out of the house and were surprised to see that their whole neighbourhood had been flattened, and roads and bridges were destroyed. Their house was saved because of the presence of long-needle pine trees in front of the house facing the sea. A boat and a jet ski, which were carried by Hurricane Ivan, could have been missiles of death and destruction on that night, but they were caught and stopped by the long and strong pine trees. Moreover, windows and doors escaped damage as they were blocked by wooden shutters. They rescued themselves from their house and moved to other accommodation, where they had to live for eleven days, mostly relying on dropped bottled water and foods from MRE (Meals Ready to Eat). Beaches and roads remain closed, and many families have lived in FEMA (Federal Emergency Management Administrator)-provided trailers for many years. The Interstate 10 Bridge across Escambia Bay collapsed.

Post-hurricane health risks take the form of trauma, injury and infectious or communicable diseases. It seems that immediate attention is usually given to water and sanitation, malnutrition, emergency shelter and medical supplies.

As Sourov was in the building trade, naturally I asked him about storm-proof buildings and underground shelters to prevent disaster. He said nowadays they were building more storm-proof houses, but underground shelters would not be applicable. However, I still feel that one day advanced technology might solve this.

Millions of people were without power. The four hurricanes cost insurance companies billions of dollars and greatly affected the tourism industry. This had a great impact on the state economy. According to some, the 2004 hurricane season might have been the costliest hurricane season ever recorded in the USA. However, this was superseded when Hurricane Katrina hit the Gulf Coast of Mexico in August 2005, followed by Rita in October 2005.

Category Four Hurricane Katrina hit the Gulf coast on 29 August 2005, killing more than 1,277 people. I was in Pondicherry, India, and watching television, which showed how the great storm was approaching the city of New Orleans from the Gulf of Mexico, and how the people of New Orleans were taking shelter in the Superdome to escape Hurricane Katrina. The irony was that, like Pondicherry, New Orleans was founded by the French in 1718 and I was watching the incoming disaster happening in an ex-French colony of the west from a post- tsunami area of an ex-French colony in the east.

New Orleans is situated on the banks of the Mississippi River, approximately 135 miles upriver from the Gulf of Mexico. Many oil rigs of the Gulf of Mexico lie just off the shore of New Orleans. Much of the city is between one and ten feet below sea level. It lies between the Mississippi River and Lake Pontchartrain, and is very prone to flooding, although protected by levees. The city has influences from French, Spanish and Afro- American cultures, and the majority of the population is Afro- American. Over the years, New Orleans has been hit by various hurricanes and has experienced several floods, but never like Hurricane Katrina, the eye of which passed within ten to fifteen miles of New Orleans.

Many oil rigs in the Gulf of Mexico were destroyed and production platforms went missing. According to the coastguard, at least twenty oil rigs were swept away or destroyed.

Heavy rains and flooding immediately followed, and the city suffered devastating effects. Levee failure was responsible for the flooding of a major portion of the city. Although predicted, one could not ignore that this was the largest civil engineering disaster in American history. 80 per cent of the city, much of which is below sea level, flooded, with water reaching a depth of twenty-five feet in some cases.

Around the world, people saw on live television the misery and horror scenes of people surviving the storm and how many old, sick and young people died. In hospital, nurses tried to revive a dying patient by hand-pump ventilator as generators and batteries failed and a million people were left without power. Food, water, power and telephone lines were gone and it was estimated that damages cost the USA $100 billion; half a million people became refugees. In the Superdome, several people died before they could get out. Hijackers, looters and gunmen tried to take over the city. Police officers battled the looters and 30,000 soldiers were deployed. Fifty-five nations all over the world offered help. The media, and many other people, posed the question: 'Was this the worst natural disaster in US history or the worst response to a disaster in US history, or both?'

After Hurricane Katrina, between 1 October and 5 October 2005, Hurricane Rita made a second assault on the Gulf Coast, with winds of 175 mph (280 km/h). The storm first struck Florida, after making an approach near Cuba, and went on to strike Texas, Louisiana and Mississippi. Parts of New Orleans were reflooded. Up to two million people fled from America's southern coastline as authorities warned that Hurricane Rita would do severe damage. People started to move from the south coast, but the movement of people was deadly slow and delayed by a one hundred-mile-long traffic jam. Petrol stations ran out of fuel, forcing police to carry petrol cans to aid motorists who ran out. It caused significant disruption to the American economy and to everyday life.

The heaviest damage was in coastal areas; however, Hurricane Rita weakened to a Category One storm with top winds of 120 km/h as it made its way inland. 23 per cent of American oil production is from Gulf Coast refineries and many were badly damaged as a result of Hurricane Rita. The reported death toll was 119 and total damage was over $9.4 billion. Over two million customers were without electricity.

On 5–6 October 2005, Hurricane Stan hit South America. It was a weak storm, hardly a Category One hurricane (80 mph or 130 km/h) but it caused flooding and mudslides, and killed 1,600 people, with a damage of over one billion US dollars. The countries affected were Guatemala, El Salvador, Mexico, Nicaragua, Honduras and Costa Rica.

2005 was the worst hurricane session in American history, and from June to October there were nine storms and twelve hurricanes. Many people blamed global warming.

Typhoons are common in the Pacific Ocean and the word 'typhoon' comes either from the legend of an Earth god or from defeng, which is the Mandarin Chinese word for 'great wind'. The strength and speed of the typhoon is measured by the rate of wind (miles or kilometres) per hour and sometimes the S-S scale is used to categorise the intensity.

Typhoon Vera in September 1959 is the worst typhoon ever to strike Japan. Its wind speed was 220 km/h (135 mph) and the disaster was intensified by the outbreak of dysentery and typhus, as a result of drinking polluted water. In 2004, Japan suffered a record of ten landed typhoons, killing 220 people with extensive damage to property.

On 12 September 2003, Typhoon Maemi hit South Korea and tore into its southern part, damaging its main port, Pusan, by knocking down cranes and sinking at least eighty-two vessels. The wind speed was up to 216 km/h (Category Four on the S-S scale), the strongest typhoon since their records began ninety-nine years ago (1904). It killed eighty-seven people and twenty-six people went missing. Most deaths were due to electrocution, landslides and drowning. Buildings were damaged and crops were ruined. 25,000 people were made homeless.

A cyclone is a large-scale circular flow, centring on an area of low pressure over the warm waters of equatorial zones and usually located in the Indian Ocean. Its strength and speed are also measured by the rate of wind (miles or kilometres) per hour. Bangladesh and India are the main countries on the Bay of Bengal that are usually blasted by tropical cyclones every year.

Some of the major cyclones that hit the Bangladesh coast occurred in 1970, when the wind speed was above 160 km/h (100 mph) and sea waves rose up to between six and fifteen metres (twenty to fifty feet) high; in 1985, when the wind speed was more than 160 km/h (100 mph); and in 1991, when the wind speed was 223 km/h (145 mph).

I particularly remember the major cyclone of November 1970, when some of the islands in the low-lying delta area of Bangladesh vanished under water. Houses were destroyed and land was submerged and flooded. 75 per cent of the staple rice crops and many fishing boats were destroyed. Many people were drowned. This great storm not only affected the coastal areas but also the mainland, and the overall death toll was difficult to estimate. According to some sources, the figure could be between 300,000 and 500,000.

Not long after this great disaster, we met an English couple at a Christmas party in the New Forest in England. I was introduced to the couple, who had been to Bangladesh (formerly East Pakistan) and had good knowledge of the Ganges–Brahmaputra Delta and its tributaries.

The wife stated, 'Most of this Bangladesh basin lies on lowland areas, like Holland. Holland solved the problem by building dykes with an efficient drainage system, and Bangladesh should follow Holland's example.'

I answered, 'Bangladesh cannot be compared with Holland. Bangladesh is a poor country. Besides the cyclone, the monsoon flood is an annual phenomenon. Moreover, the rivers of Bangladesh often change their course, resulting in devastating effects. The preventive measures taken by Holland are not that easy for Bangladesh to develop.'

A year later, I went to India and met some people from Bangladesh who had escaped death in this disaster by tying themselves to trees and taking shelter in the bush. It indicates the importance of forestation.

The last violent cyclone from the Bay of Bengal that struck the eastern cost of India, mainly the state of Orissa, was in October 1999. With a wind speed of 160 mph, a thirty-foot-high tidal wave had the devastating effect of uprooting trees, flattening houses, flooding coastlines and washing away villages. According to official estimates, twelve million people and 7,921 villages were affected by this cyclone. It cost 9,615 human lives and the deaths of 400,000 cattle. 800,000 houses were damaged. Many fishermen did not return home. Emergency rescue teams were set up and doctors, the army and other emergency services were deployed for the rescue operation and for relief work.

This reminds me of my first involvement in such a major disaster, in 1967. I had just finished my final medical examination and the Indian Medical Association wanted to send a team of doctors to the cyclone-affected area of the West Bengal coast. So I volunteered. The high wind and high tidal waves from the Bay of Bengal affected the coastal districts of West Bengal and Orissa, and what was then East Pakistan (now Bangladesh). The coastal area of the Midnapur district was heavily flooded and there was urgent need for rescue and relief work. We were a team of three doctors and the first medics to arrive, geared up with medicine and tablets, vaccines and immunisation tools and equipment, ready to go to the flood-affected area, which was approximately 185 miles south of Calcutta. We were told that, when we arrived there, we would get further support from local medical practitioners and other government officials.

We took a train from Calcutta, and arrived at Kharagphur railway station, which has the longest platform in the world, measuring 2,733 feet in length. From the station, we took a minibus and reached our first destination, midway between Kharagphur and a small seaside resort town, Digha. It was nearly mid-morning, and the place we reached was the Block Development Office, situated near a local canal. The Block Development Officer (BDO) greeted us. We discussed the areas and the villages affected, and access to those places. The public health issue was a big topic. The sanitary inspectors reported to him. To prevent epidemics like typhoid, cholera, smallpox and malaria, vaccination, immunisation, bleaching powder, DDT, non-polluted water and tablets were all important items. As DDT was a cheap and effective insecticide, it had worldwide uses, especially for malaria control. DDT, whose full name is dichlorodiphenyl- trichoroethane, was invented by the Swiss chemist Paul Herman Mueller (1899–1965) on the eve of the Second World War. He won the Nobel Prize for Physiology and Medicine in 1948. Unfortunately, DDT can contaminate the soil, mammals, birds, fish, and even people through food consumption. Due to its increasing toxicity, most of the countries of the world agreed to control its use by banning or severely restricting its use. India and China are the only countries that still produce DDT.

Smallpox used to be a big health issue but it is now completely eradicated from the world. However, one cannot rule out its reappearance in the natural cycle, or as part of chemical or biological warfare or terrorist attack.

The BDO informed us that our next journey would be through the canal by country boat, and he arranged the boat and guide. In the afternoon, our boat journey began for the next destination.

On the way, we saw that the dykes on the banks of the canal were intact, although in some places paddy fields were flooded. We saw very few houses on the riverbanks and occasionally passed through unremarkable villages and noticed small bushes, including bamboo bushes.

It was late afternoon, when we suddenly heard the sound of a bell from the bushes and a man appeared in khaki dress with a head turban and a sack on his back. He was running or walking very fast, slightly leaning on his tall stick, which he was holding with his right hand; the bell was attached to it. He disappeared as quickly as he appeared. However, we could still hear the sounds of the bell from the boat, which gradually faded away. It happened twice; the first time, we could not make out who he was. Was he a night watchman or a porter? The second time, we were able to identify him as a post runner or foot messenger, who carried the post from one village to another; they usually worked for the post office. I had never seen a post runner before and it was a great excitement for me. He was doing a very good job, carrying the mail through flood-affected areas. I had thought that, with the improvement of transport and communication systems in India, most of the post runner positions were gone or abolished. However, there were remote places in India where there was no proper access and the only way to reach those places was by post runner. That was more than forty-two years ago, but, even in this century, it would not surprise me to see post runners in some parts of India, carrying on the same old traditional duties.

After two hours on the boat, we reached a very narrow canal, and our boatman told us that the boat could not go beyond that point and that we had to go to the next destination on foot. We got out of the boat, picked up our medical supplies and our personal luggage and walked through the chest-deep water. Our medicine boxes were safe, as our boatman and his assistant carried those boxes on their head.

It was evening when we reached a small health centre. This health centre was without any doctor, and we were told that the place was so remote that nobody wanted to come here. In India, it was not uncommon that sometimes health centres and hospitals were built for political and financial reasons without priority given to the suitability of the location. Our appearance or presence for relief work was suspicious to some local people. The local political leader, I was told, belonged to the state non-ruling party. He was a little hostile and asked us, 'What political group do you belong to?' We explained, 'We are not representing or belonging to any political party. We are from the Indian Medical Association and are volunteering for the flood relief work.'

It was interesting that the health centre was situated on high ground. So it was not affected by the floodwater, although some of the surrounding villages were submerged.

We travelled quite a lot by boat or on foot; occasionally on bike or in a Land Rover. We saw how people were marooned and how some took shelter on the high ground, in treetops and on the roofs of houses. I also saw a man hanging from a coconut tree. Many houses were destroyed, especially those built of mud. Many paddy fields were flooded and crops destroyed.

The rescue process was slow, until the army with their speed-boat came to the rescue. We saw some places where the sea dyke had broken and pine woodlands that provided protection from the wind had been damaged by sea flooding. The cashew plantation is quite famous in this part of the coast, and fortunately the damage was less here. We managed to taste some raw cashews, although there were very few nuts on the trees. However, we were able to taste some delicious sweets made of cashew nuts, in the local market of a place called Contai, a sub-divisional town of Midnapur district. The place had been very little affected by the cyclone and flood.

We stayed in all kinds of accommodation, including beautiful mud houses, and covered as many as villages as possible, treating flood victims and providing immunisation. I spent two and half weeks there, and it was indeed a great experience.

On the last day, I was in a small seaside town, Digha, the place that was least affected. We saw a private aircraft on Digha beach, which was intact, and it seemed that the aircraft had escaped the cyclone. This aircraft belonged to the chairman of a renowned multinational company. His office was in Calcutta and every weekend he used to fly from Calcutta to this small resort, where he had a private holiday home. Landing and taking off from the long stretch of his private sandy beach was routine to him.

The 2 May 2008 was the day of the worst natural disaster faced by the Burma (Myanmar) people. It was a severe tropical cyclone, named cyclone Nargis, originated in the central Bay of Bengal on 27th April 2008; it landed on 2 May, causing devastation to southern Myanmar. It was originally 215km/h (135mph) storm which became 165km/h (105mph), when it made landfall on the Irrawaddy delta. Ninety-five percent homes were destroyed, many pagodas were damaged and fields, roads, ditches, trees, houses were washed away as a result of strong winds along with twelve feet high waves. Sewage mains burst and saline water along with sewage was responsible for the destruction of entire paddy fields. According to the United Nation 2.4 million people were affected, with 55,917 missing and 77,738 dead, although some estimated that the final toll would be more than 100,000. Refuges, disease, starvation and looting were big issues. Most of the children suffered from diarrhoea, and there was an acute risk of malaria and typhoid epidemics. The skins of many victims were lacerated due to the powerful wind.

The UN world food programme and WHO medical supplies were not able to operate before 21 and 19 May respectively. The Disasters Emergency committee representing thirteen major charities found that they were not able to reach many areas. Anger developed amongst survivors because of delay in aid. Many deaths could have been prevented if aid had been accessed in time. Initially the Myanmar government was reluctant to accept international help, but later it agreed to offers of help. Forty-six countries came forward with aid in the form of food, medicine, water, shelter materials, cloths, blanket, mosquito nets, emergency relief equipments and medical personal. Mini hospitals were set up in the Irrawaddy delta. The Pandav ship on which I cruised on the Ganges in autumn 2009 was used as a hospital ship to provide medical services to Irrawaddy delta affected by cyclone Nargis.

Tropical cyclones are a constant threat to the Northern Territory of Australia, and the Australia Bureau of Meteorology has categorised cyclones on a scale of one to five. Category One is less than 125 km/h with negligible damage, and Category Five causes extreme damage with a wind speed of more than 280 km/h .Category Five is extremely dangerous, causing widespread destruction. The Australian Cyclone Tracy, with wind speeds of 217 km/h (135 mph) struck Darwin over Christmas 1974. It was the worst ever storm in Australian history. Darwin became a ghost town as a result of that five-hour storm and the destruction, death and mass evacuation it caused. Many people fled or left their homes; the injured and homeless were evacuated.

Strong winds can blow seawater up rivers and can cause floods. A major example is the flood on the Thames in 1953, when a bad storm pushed seawater into the mouth of the Thames, causing dangerous flooding and affecting 160,000 acres of farmland and 200 miles of railway lines. About 300 people, 56,470 poultry and animal stocks drowned. 24,000 houses and 214 major industrial premises, including two power plants and twelve gas works, were affected or damaged. The River Thames begins its 346-kilometre (215-mile) journey about a mile north of the village of Kemble (in the Cotswold Hills), near Cirencester in Gloucestershire, and passes through beautiful villages, towns and cities, ending in the North Sea. The River Thames famously flows through London and is the longest river in England. It has a long history of flooding. To prevent this – and the possible inundation of London – the Thames Flood Barrier was built across the river at Woolwich, and has been in operation since 1982. I have visited it a couple of times, while cruising on the River Thames. The Thames Barrier, which is about 529 metres in width from one bank to the other, is a tidal barrier that prevents the sea water coming in on high tides. However, for the past few years, in order to prevent flooding, the barrier has had to be used more often than usual.

A tornado is formed in inland areas as a thundercloud, and this updraft of warm air is then affected by winds at different speeds. This causes air to rotate; a tornado is formed by strong winds rotating in an upward spiral column, with a lighter descending wind at the centre.

When the column touches the ground, it generates suction that can rip up trees, tear off roofs and carry heavy objects skywards.

High to low tornadoes are a global event and are seen when warm, moist air clashes with cool dry air. The USA suffers more tornadoes than any other country. Every year, warm, moist air from the Gulf of Mexico clashes with cool, dry air off the Rocky Mountains. Tornadoes (or 'twisters') take place in the spring and early summer, and mid-west America has the greatest incidence of injury, death and damaged property. Tornadoes are responsible for $400 million of damage each year in the USA. The intensity of damage is measured on the basis of the Fujita Scale, which is determined after examination of radar tacking, eyewitness testimonies and the damage resulting from the tornado. The scale contains six categories, in order of increasing intensity (Table 3).

Table 3: Fujita Scale

Category	Wind speed	Damage
F0	<73 mph (<116 km/h)	Light
F1	73–112 mph (117–181 km/h)	Moderate
F2	113–157 mph (182–253 km/h)	Significant
F3	158–206 mph (254–332 km/h)	Severe
F4	207–260 mph (333–419 km/h)	Devastating
F5	261–318 mph (420–512 km/h)	Incredible

The most violent and destructive tornado of the last century took place on 3 May 1999 in Oklahoma City. It was F5, left incredible damage which cost $1.2 billion and killed forty-two people. This was the first time in US history that tornado damage had exceeded one billion dollars.

In this century, there have been eighteen major significant tornadoes or tornado outbreaks around the world, but none was F5 category; the highest was F4.

The single deadliest tornado was on 26 April 1989, in Bangladesh, which claimed 1,300 lives.

Personally, I have come across twisters in India, but not in the USA. Twisters are usually seen during the heat and dust of the Indian summer in the months of April and May, and are mostly associated with summer or tropical storms. The warm, moist air from the Bay of Bengal collides with the cool, dry air of the Himalayas. In my school days, I twice got caught in a mild twister. On one occasion – I remember it was late afternoon – I was crossing the road and I suddenly saw a column of dust twisting and rotating upwards in front of me. It travelled away very quickly, leaving dust particles that got in my eyes, and I had to struggle to get them out with a corner of a handkerchief. On another occasion, a twister carried my exercise book away; I ran after it and was able to rescue my book after chasing it for a quarter of a mile. I was not sure whether the second time I did it for fun or not; it was certainly a great experience.

As I have described, storms, winds and high tidal waves cause death and destruction. Fortunately, nowadays such destruction is less, due to early evacuation. Rich countries are able to undertake the appropriate preventive measures prior to the storm striking the predicted areas. This is possible because of the advancement of accurate weather prediction and fast media technology.

However, more research is needed for a device that will modify hurricanes by weakening their winds or diverting them from populated areas. More emphasis is to be given to more effective storm-proof buildings.

The future question is how to utilise this wind power scientifically in more productive ways. Already some countries are generating electricity by utilising coastal winds and high tidal waves.

Onshore wind power has been criticised by many because wind turbines are not very effective, are noisy and are not friendly to wildlife. However, an appropriately designed wind turbine which is less noisy, friendly to wildlife and more efficient will solve some of the problems. Britain is planning to develop wind power further by building wind turbines on the seabed in certain coastal sites within five miles of the shore. Already, electricity has been generated successfully for the past three years from wind turbines on the seabed. The biggest one is located off the coast of north Wales, and on a clear day it is easily visible from land. In the mean time in September 2010, the world's largest offshore wind farm project opened in the south-eastern coast of Britain. This is located about 20 km (12 miles) from Kent and Essex coast, spreading 35 Square km. When it is completed it is expected to generate 1000 megawatts of electricity by the hundreds of seabed wind turbines..

This way, some of the clean energy problems of the world will be solved, but this type of project is expensive and difficult to implement in some parts of the world, where cyclones, hurricanes and typhoons are common, if not annual, occurrences. However, if turbines can withstand such hostile environments, then there will be a great scientific breakthrough. It will benefit the countries that lie within those disaster zones.

Chapter Six

Avalanches, Mudslides and Landslides

Landslides

Over the past few decades, the death toll through avalanches, mudslides and landslides has increased throughout the world. More attention is therefore needed from scientists, researchers and rescuers to act on those disasters. This is mainly because we are experiencing an increase in activity on mountains and hillsides, but climate change, global warming and the erosion of lands by rivers, glaciers and ocean waves are other contributing factors.

An avalanche is a mass of snow that slides down a slope. It has a distinct path that consists of a starting zone, track and run-out zone. The starting zone is higher up on the slopes; the track is the channel that an avalanche follows as it goes downhill; and the run-out zone is where the snow finally comes to a stop. In the snow, there are debris, rocks, and vegetation which follow the same pathway and are piled or deposited on the deposition zone. Thus, in descending, an avalanche may grow to an enormous size by picking up additional snow, rock and debris. The deposition zone is where victims are usually buried. Avalanches are sudden, violent and destructive, although the conditions of the avalanche might vary due to certain factors, and these could be responsible for creating low, moderate or severe avalanche conditions. These factors include temperature, storm and wind direction, snowfall, slope, terrain, vegetation and the stability of the snow pack.

For centuries, mountain-dwellers and travellers have been victims of avalanches but, for the past few decades, mountain climbers, trackers, skiers and tourists have been caught more often in avalanche disasters. Recreation and sports activities attract millions of people towards mountains. The mountain industry is booming. The great examples are the Alps, the Himalayas, and the Colorado and Rocky Mountains. Winter sport-related avalanche danger areas are mainly in Europe, Canada and the USA.

The Andes is not famous for winter-related recreation or sport activities, as there is no mass snow and it is mainly a semi-arid area. I noticed this when we visited Peru and Bolivia during winter (July 2007), even though we were between 13,000 and 14,000 feet of altitude. The only snow we saw was at a distance, on the peaks of mountains. This is because of close proximity to the equator. Therefore, travellers are not victims of avalanches, but many do suffer from altitude sickness.

Every year, thousands of people from all over the world visit Cusco, Machu Picchu and Lake Titicaca. Visitors, tourists and trekkers can suffer from this sickness, and we were not the exceptions, as we travelled at altitudes of between 2,340 metres (7,970 feet) and 4,313 metres (14,108 feet). This form of sickness may develop at altitudes greater than 2,000 metres (6,500 feet) and is due to the lack of oxygen at high altitudes. However, I was surprised to see how the people of these areas are able to play international football games, even at such a high altitude. The risk factors of altitude sickness for visitors are rapid ascent, higher altitudes and greater exertion. People living in low altitude areas and those with a past history of altitude sickness are more vulnerable.

The symptoms are usually headache, dizziness, nausea, vomiting, loss of appetite, malaise, insomnia and breathlessness. Severe cases may develop high altitude pulmonary oedema (fluid in the lungs) or high altitude cerebral oedema (swelling of the brain). Oxygen inhalation and descent to lower altitudes improve the condition. People may use Diamox (acetazolamide) or dexamethasone to counteract the symptoms.

With this knowledge, I tried to evaluate the effect of altitude sickness among our fellow travellers. There were thirty people who volunteered, of which seven were from the Netherlands, one from Italy, and the remaining twenty-two from the UK. Of the twenty-two British residents, fifteen were from England, four from Wales, two from Scotland and one from Northern Ireland.

Overall, there were fifteen males and fifteen females. Seven were below fifty years of age, and the remaining twenty-three were above this age. Twenty-three people complained of some altitude sickness during their six-day stay at high altitude. In the below- fifty age group, six people reported symptoms, but in the above- fifty age group, only six people did not complain of any symptoms. The symptoms were headache, feelings of light-headedness, nausea, vomiting, nosebleeds, dizziness, shortness of breath, irregular heart and pulse rate, weakness and lethargy, muscle pain and flu-like sickness, eye problems, stomach pains, bloated stomach and lazy or slow bowel-related issues. Some symptoms may have been due to other health issues, and one person complained of side-effects from Diamox. Two people needed oxygen and one person required specialist treatment. Oxygen is commonly available in most hotels and travel coaches in these areas. Most people took the local coca tea, which is known to reduce altitude sickness. Seven people were on Diamox tablets. As the Netherlands is below sea level, it was expected that the people from that country would be vulnerable to altitude sickness. Of the seven, all except one had altitude sickness. None from the group reported a past history of this condition. I also noticed that most of us struggled to climb steps and uphill areas. This was obvious when compared to the local people, who were walking and climbing at a fast pace without any signs of difficulty. However, in the end we all recovered gradually when we came down to the lower plains.

In the Himalayas, since the first successful ascent of Mount Everest on 29 May 1953 by Sir Edmund Hillary and Tenzing Norgay, 1,200 men and women have reached the summit. Each year, more than 20,000 tourists visit the Khumbu Valley beneath Mount Everest, the tallest peak on Earth. Tenzing Norgay (1914– 1986) is dead now, but I met him more than forty-five years ago at the Himalayan Mountaineering Institute, located in Darjeeling, West Bengal, India, where he became the first director of field training. Sir Edmund Hillary (born 1919) died at the age of eighty-eight in his native country of New Zealand on 11 January 2008. The irony is that Tenzing Norgay died in Darjeeling and I was visiting this very region when I heard the news of Sir Edmund Hillary's death.

The construction of more roads, buildings and towns is on going to keep pace with the increased demand for mountain activities worldwide, encroaching on avalanche-prone areas.

Avalanches swallow up many climbers, skiers and snowboarders. Each year, avalanches claim up to 200 lives worldwide, and thousands more are caught up in the form of injuries or partial burial. These numbers are always increasing. France has the highest number of avalanche deaths. Melting snow is the most common reason for avalanches taking place.

Before one plans to visit an avalanche-prone area, one should have proper health and safety education. Some countries now run an avalanche-awareness course, which might be helpful. Carrying a mobile phone, shovel, air bag are also useful. One needs to be aware of what to do if one is caught in an avalanche, how to survive an avalanche and how to rescue a victim.

In the event of an avalanche, when a colleague or friend is the victim and is carried down a slope and buried under snow, it is important to look for clues and search the area. The role of trained dogs as a part of avalanche search teams is no doubt extremely valuable in this respect. Once the victim is found, dig as quickly as possible, as the survival chances decrease rapidly depending on the duration of the burial. If the victim has survived, one might have to treat him or her for shock, hypothermia or injuries.

The most common causes of death in an avalanche are suffocation, the impact of the snow mass and hypothermia.

A severe avalanche is not a new phenomenon in the Swiss Alps. A major avalanche with a speed of 364 km/h (215 mph), covering 6.9 kilometres (4.3 miles) in seventy-two seconds, has been recorded in Switzerland. 1999 was the worst recorded winter in central Europe, with the heaviest snowfall seen in fifty years. Tens of thousands of tourists were trapped in the ski resorts. Two avalanches, one 200 metres wide and another forty-five metres deep, hit a ski resort and a nearby village in western Austria, killing seventy-five people during January and February 1999. The houses of the Austrian villages Galtür and Valzur were destroyed by massive snow slides; the majority of the deaths occurred 3,000 metres up in the mountains between Ischgl and Galtür. There is a memorial in Galtür for the victims of the 1999 avalanches. I also learnt that, since then, to protect the village from snow slides, Galtür has constructed a protective wall and shelters.

Avalanches had been frequent after the heavy winter of 2009-2010. There were twenty-eight deaths in the Swiss Alps alone, somewhat higher than average. Ninety-five per cent of the avalanches were triggered by skiers and boarders, although there were high avalanche danger warnings and in some places, it was rated three or four on the five-level hazard scale rating. Heavy rain and then snow had fallen on weak layers in the snowpack causing the snow to become very unstable and leading to the hazardous outcome.

It is well known that avalanches can be triggered by artillery fire, bombs and blasts. The great example is the two World Wars.

In the First World War, about 40,000 soldiers perished beneath the snow because of avalanches triggered by Italian and Austrian gunners.

More recently, the regular exchange of artillery fire between Indian and Pakistani soldiers on the Siachen Glacier, in Kashmir, is often blamed for triggering avalanches, killing both Indian and Pakistani soldiers. Indian and Pakistani soldiers have been fighting each other in harsh conditions for last twenty years on this roof of the world, ranging from 16,000 to 20,000 feet; this is the world's highest battleground. However, for the past five years there has been an ongoing ceasefire between India and Pakistan. The Siachen Glacier, seventy-two kilometres long, is one of the longest glaciers on the earth, lying in the East Karakoram and West Himalayan mountain chains.

Most glaciers are found in mountain areas. The snow collects in the hollows, high on the mountain, and remains in the same area, year after year, transforming into compact ice as its weight builds up. A glacier is a great mass of ice that spills out from permanent snowfields and gradually slides downhill.

When I visited the Greenland in August 2010, I saw how the ice accumulates on the ice sheet and then melts or calves out from the glacier, produces icebergs and reaches the sea from the Sermeq Kujalleq (Ilulissat glacier) and the Equip Sermia (Eqi glacier), the most popular glacier outlets in Greenland. The most possible cause of calving is the tendency of the ice to spread out.

In a warm climate, the glacier melts, and rivers drain the glacier and its water into the sea. On the way, it sometime causes inland flooding or creates a lake through natural damming of the water from the melting glacier. Kashmir and Ladakh have many such glaciers, and visits to those places are really a great adventure.

We were lucky to pay a visit to one of the glaciers, as, since 1989, it has been very difficult to visit Kashmir due to increased militant activities. However, we visited the Sonamarg-Thajiwas Glacier in Kashmir in 1986.

Sonamarg is located at a height of 9,000 feet (2,740 metres), fifty-one miles (eighty-one kilometres) from Srinagar, and this is the last major place in the Kashmir valley before entering Ladakh through Zojila Pass. From Srinagar, our journey to Sonamarg was by car. On our way, we passed through Sind Valley, which is drained by the Sind River.

The landscape changed from green to barren as we approached the glacier. The sight of some parts of the glacier appeared to me as a receding glacier. With the heavy snowfall, from time to time in winter, avalanches sweep down the glacier and, with the summer heat, the glacier melts. Glaciers and melted water are the major source of erosion in barren areas, since they have little vegetative cover. A receding glacier leaves behind unconsolidated debris that may, in the course of time, become the source material for future landslides. On the way to Sonamarg and on the return journey to Srinagar, we did not see any landslides or mudslides, but noticed some flash floods along the road and the fields, more towards Srinagar. We also saw some shepherd boys crossing the roads with mountain goats.

Over the next couple of days, our plan to visit Pahalgam was totally ruined due to unexpected mudslides and landslides. They occurred without warning. Pahalgam is ninety-five miles away from Srinagar and is situated at the junction of the Aru and Sheshnag Rivers at an altitude of 2,130 metres. The Aru flows from the Kolahoi Glacier, and the Sheshnag flows from glaciers along the Himalayas.

We started in the morning from Srinagar by coach and, after passing Anantnag, the coach took the route north-east to Pahalgam. After travelling a few miles, the coach suddenly stopped and driver told us that the couch could not proceed further as the road was blocked due to a landslide or mudflow.

I got down from the coach and walked half a mile down the hill for a glimpse of the incident site. When I reached the spot, I was surprised by the nature of the destruction. I had seen a landslide before but this was beyond my expectations. It was a combination of landslide, mudslide and flood. The tarmac road was completely destroyed on that particular spot and there was a big ditch, the entire width of the road, from right to left. On the right side of the road was the hill and on the left side was the valley. At that particular spot, water was coming down the hill, carrying debris, mud, rock and vegetation across the road and through a five-foot-wide ditch leading towards the valley. I had never before seen the speed and the way in which water flowed through a newly created channel.

I recorded it on my video camera. Some of the Kolahoi and Sheshnag Glaciers must have melted. There was no doubt that this road to Pahalgam was completely cut off. There was no other alternative but to return to Srinagar and to abandon our trip to Pahalgam.

A few days later, we were again halted by landslides. The road was blocked and we were trapped at the Jawahar Tunnel. This was the day we were leaving the Kashmir Valley by car. The Jawahar Tunnel is on the national highway linking the Kashmir Valley and Jammu, from where one has to catch to train to go anywhere in India. The tunnel is two and a half kilometres long and is located on a treacherous mountainous road ninety-three kilometres from Srinagar. Before the tunnel was built,

Srinagar was completely cut off in winter from the rest of India, and the only approachable road was through Pakistan. We were stranded at the Jawahar Tunnel, along with hundreds of other motorists. We were worried we would miss the train. It took four or five hours to clear the landslides. Eventually, the traffic started to move and so, fortunately, we were able to reach Jammu in time to catch our train.

Our trip to Kashmir was eventful. The natural calamity and also the diarrhoea that affected my health did have some impact on our tour. I had never had such a bad experience in my life. Despite taking all necessary precautions, I was having constant runs. I was puzzled whether this was hill diarrhoea or due to polluted water or a kind of food poisoning, although I had been particular as to what I drank, how I drank and what I ate. In my opinion, the hotel in which we stayed was possibly responsible for my stomach upset, though my fellow travellers did not agree with me. They were my wife, my two daughters and Bikram Mohanty, a non-medical colleague of mine from Rourkela. Bikram is now vice president of a Hyderabad-based Indian company.

The pollution in Srinagar was a great issue. I was sure that the River Jelhum did not maintain a good quality of water and did not fulfilling the criteria for drinking water. Moreover, when we took a boat trip on the Dal Lake, I noticed that it was polluted and sinking due to waste and human encroachment along the bank. I was glad that in 1986 the state government of Jammu and Kashmir constituted the State Pollution Control Board to prevent pollution under the Water Pollution Act, 1974 and the Air Pollution Act, 1981.

However, the hospitality we received from the local people was a great experience.

Mr Warr, who was a local Kashmiri businessman, took us to various interesting places in Srinagar. We met some local people and saw local crafts. Kashmir is famous for walnuts and for furniture made of walnut trees. We saw such beautiful furniture. Mr Warr also invited us to his home, located in the modern part of the city. We met his entire family and had a splendid lunch in his house in the local manner. We enjoyed Kashmiri rice, which is famous in Kashmir, especially the way they prepare it. From him, I learned that Kashmiris do not eat beef. It might be true for the Kashmiri Hindus or Pundits, but it was difficult to believe that the Muslims of the Kashmir Valley do not eat beef. I certainly had not noticed it when we had eaten a meal with a Kashmiri Muslim in England. Mr Warr also told me that outsiders cannot buy property or build houses in the Kashmir Valley, and this tradition has been ongoing since the period of the British Raj. The British could not build houses or buy property, and so instead lived in Dal Lake houseboats and enjoyed the cool climate of the Kashmir Valley in summer, away from the dust, heat and humidity of the Indian plain. Mr Warr had three daughters, and one daughter was getting married soon. We had a splendid time.

Landslides are serious geological hazards and occur due to a wide range of ground movements, such as rock falls, deep failure of slopes and sallow debris flow. The primary reason for landslides is gravity acting on a steep slope. Other contributing factors include overly steep slopes created as a result of erosion by rivers, glaciers or ocean waves; weakened rock and soil slopes through saturation by heavy rains or snowmelt; excess weight from the accumulation of rain or snow, stockpiling rock or ore, waste piles or other man- made situations (even buildings can put stress on a weak hillside); deforestation and irregular logging; typhoons and storms; earthquakes; and volcanic eruption.

Landslides are, therefore, the dislodging of matter from hillsides that cause a mass of earth and rock to fall. A mudslide or mudflow is one type of landslide but the difference is the presence of water, which is usually lacking in landslides. A mudflow is therefore a flow of water that contains large amounts of suspended particles and silt. Steep hills and mountains are often the sites of landslides and mudslides, and mudflow occurs on slopes. The topographical causes of mudflow and landslides are a combination of the slopes of hills and mountains that have become weakened by erosion, fires, heavy rains, snowmelt, earthquake and even buildings. When mudslides and mudflow occur, homes and property are buried or swept away. People are left homeless, injured or killed. Thousands of people worldwide die every year from landslides and mudslides.

In August 2010, the mudslides that occurred in the north-west of China following thunder, lightning, storms and floods, were responsible for a valley disappearing under the mud and killing 127 people. 1,294 people were reported missing and 300 homes were buried.

Landslides are not common in Britain, although from time to time they happen. A notable example is the landslide in Scarborough in 1993. When we were in Teeside, we used to go often to the seaside town of Scarborough for a break. I remember a nice beach, the cable car, the cliff top and a hotel on the top of the cliff, named Holbeck Hall. This hall fell into the waves as the cliff collapsed. The landslide was triggered as a result of undercutting wave action on the cliff over the years.

Landslides are most common in Japan, as 70 per cent of the country is mountainous and hilly. In the rainy season, mudslides occur in Indonesia, Malaysia and across south-east Asia.

Typhoons and storms hit the Philippines every year. In November 1991, about 6,000 people were killed in floods and landslides triggered by a storm. In early December 2004, 1,000 people died or went missing and there were 30,000 victims in total as a result of the floods, landslides due to typhoons and the three big storms which hit the Philippines. The blame for this tragedy has been placed on years of over-logging in nearby hills and mountains.

In February 2006, there was again big news: 'Philippines mudslides kill 1,800 people'. It happened in the central Philippines island of Leyte. Storms and weeks of heavy rain caused the collapse of the mountainside, which engulfed a farming village, Sogod, in Guinsaugon, leaving more than 1,000 people buried alive under metres of mud.

In 1990, we suddenly heard the sad news that two multi- storey buildings, located on a famous hillside in Kuala Lumpur, had collapsed as a result of mudslides. Most people inside were trapped and, except for one child, all were killed. Among the victims were one of my wife's cousins and his wife. The building did not withstand the heavy rains. It was difficult to identify the bodies. Parts of the bodies of both the husband and the wife were found on the stairs holding each other's hands.

Landslides, mudslides and mudflows leading to the collapse of buildings are often caused by mismanagement and unwise land use practice on grounds of stability. Planning and development regulations, construction codes and standards, and vulnerability assessments are all important to prevent such tragic incidents. Likewise, more importance is to be given to the complete banning of irregular logging, especially in a country like the Philippines where it is a routine phenomenon.

This shows the importance of risk assessment, risk management and public awareness. More focus is needed for effective warning, danger scale ratings and their colour-coding, quick checks and extra precautions, including protective measures or devices. Lastly, there must be effective education for the prevention of avalanches, landslides and mudslides.

Avalanche information and danger rating levels vary from country to country and even within countries. Recently, after considerable debate, the five-level Unified Risk Scale has been adopted by the majority of countries in the world.

Unified Risk/Hazard Scale

1. Low

Well bonded, stable snow pack.
Triggering is possible only with high additional loads* on a few very steep extreme slopes.
Only a few small avalanches are possible.

2. Moderate

The snowpack is well bonded generally, but moderately bonded on some steep slopes.
Triggering is possible with high additional loads* particularly on the steep slopes indicated in the bulletin.
Large natural avalanches are most unlikely.

3. Considerable

The snowpack is weakly to moderately bonded on many steep slopes.
Triggering is possible, sometimes even with low additional loads.*
The bulletin may indicate many slopes that are particularly affected.
In certain conditions, medium- and occasionally large- sized natural avalanches may occur.

4. High

Weakly bonded snowpack in most places.
Triggering is possible even with low additional loads* on many steep slopes.
In some conditions, frequent medium- or large-sized avalanches are likely.

5. Very High

Generally weakly bonded and largely unstable snowpack.
Numerous large natural avalanches are likely, even on moderately steep terrain.

*Additional load:

High: group of walkers, climbers, skiers.

Low: individual walker, climber, skier.

In future, significant importance has to be given to geotechnology. Effective utilisation of scientific understanding must be made regarding the causes of landslides, mudslides, mudflows and avalanches.

Eqi Glacier, Greenland

Chapter Seven

Gales, Snow and Hail

Frozen Britain, January 2010

The world is usually divided into five major climate zones, namely: polar, cold, temperate, dry and tropical. The polar climate is the tundra climate – the areas mostly covered with snow, treeless, an icy land mass. The cold climate has long, very cold winters, snow and short, cool summers. A temperate climate occurs in humid middle latitude areas with warm, dry summers and cool, wet winters. Dry climates are mostly desert (arid) or semi-desert (semi-arid) areas where rainfall is very little and there is a high daily temperature. The tropical climate has a high temperature, humidity and heavy rains. These areas are mostly rainforest and savannahs.

I was born and brought up in a tropical climate, and I did not have any experiences of gales and snow until I arrived in Scotland in my early twenties. Prior to that, the only snow I had seen was on Himalayan peaks when I was spending my summer holidays in hill stations like Simla or Darjeeling in India – and that was only as a tourist attraction to see a snow-peaked mountain from a distance. Kanchanjunga was one of the spots which could be viewed on a nice sunny day from the Tiger hill, Darjeeling.

Kanchanjunga, which is more than 28,000 feet high, is the third-highest peak in the world, situated on the Sikim–Tibet border in the Himalayan range. I had a spectacular close view even from our hotel window when I visited Sikim in later years.

However, I had a little glimpse of snow and snowfall in January 1968, when my flight stopped at Zurich airport in Switzerland.

Snow is made when water condenses at temperatures below freezing, and its density is 0.05 grams per cubic centimetre. It is formed from minute ice crystals and, in very low temperatures, these crystals fall to the ground as a fine dust. As the temperature warms up, the crystals stick together to make snowflakes. Single snowflakes join together to make large snowflakes, indicating that the air is cold enough for the snow to fall.

A gale is a strong wind, and blizzards are the combination of falling snow and strong winds with poor visibility.

My first experience of snow and gale was in Ayrshire, Scotland. I remember, as a junior doctor, I had just finished my duties in the ward and was planning to go out. However, I could not, because the snow and gales were creating dangerous conditions. I looked outside through the window and saw the snow falling. It was a winter evening and there was snow all over. It was shining like a diamond as the streetlights reflected on the snow crystals. There is no doubt that falling snow is a wonderful sight, but it can become deadly, especially in blizzard conditions. Some of the nurses were worried about going home. We started to chat about snow and gales and one of the nurses asked me whether I had ever seen snow in Calcutta. I replied that I had. She was surprised, as everybody knows that the temperature in tropical regions does not fall low enough to enable snow to fall. I explained: 'Hailstones.' Lots of people do not know what is meant by hail and hailstorms. Hail is a kind of icefall. It usually happens after the summer heat in a tropical country, where the raindrops in a thunder cloud, instead of falling, may be swept up into the cloud a number of times. The raindrop freezes and becomes an ice pellet. The ice pellet grows larger each time it goes upwards. Finally, it is so heavy that it falls as a hailstone.

Hailstorms happen suddenly and are over quickly. Hailstones are sometimes so big and so sudden that they can kill people. Every year throughout the world, hailstorms not only damage crops and property but also injure and kill humans and animals. Crop damage has serious implications for agriculture, especially in rich countries like the USA, which spends billions of dollars every year on crop damage.

There have been incidents where hailstones have broken the windshields of cars. A huge hailstone once made a hole in a plane, resulting in a frightening experience for more than 200 passengers.

The stones can be as small as a pea (half an inch in diameter) and as large as a baseball (two and three quarter inches in diameter). The majority of hail is small, but large hailstones are not uncommon. According to the Guinness Book of Records, the heaviest hailstones, weighing one kilogramme (two pounds, three ounces), were reported from the Gopalganj district of Bangladesh on 14 April 1986 and killed ninety-two people. In 1975, a heavy hailstorm was responsible for multiple crashes and only six cars are left on the circuit when the British grand prix race finishes. The worst hailstorm ever recorded killed 246 people in Moradabad, Uttar Pradesh, India, on 20 April 1888.

I can recall my first experience of hailstones, and this was during my early childhood in Calcutta. After the hot, humid summer, everybody was looking to the sky for rain, as there were dark clouds gathering. Suddenly, I could hear a heavy shower of stones, hitting windows and the roof of the house. I went up to the rooftop and saw the hailstones, which stopped as quickly as they appeared. Within a few seconds, they had covered most of the flat rooftop with ice cubes. It is interesting to mention that I used to suffer from prickly heat and one of our next-door neighbours who was there rubbed my prickly heat-affected area with hailstones and told me this would help to cure it. I am not sure whether it had some therapeutic effects or not, but it certainly helped to cool down the affected areas. For a number of years now, I have been away from India and my visits to Kolkata have been short, so I have not seen any hailstones for a long time.

My understanding of a gale is of a bitterly cold northerly wind, which originates from north of the British Isles (that is, from the Arctic or North Pole), and according to the Oxford English Dictionary a 'gale' is a strong wind that usually measures 32– 54 mph.

Once, we were caught in a gale when we were crossing the English Channel. Our small ferryboat was thrown up and down in high waves and a northerly wind. Many passengers were sick and some sustained minor injuries. The captain requested extra help from passengers who were medical doctors. I volunteered, but there were a few more medical volunteers who were also there.

Gales are probably one of the reasons why I did not take up golf as a hobby. When I was working in Scotland in the late 1960s, one of my Scottish doctor colleagues took me to the golf course at Troon in Ayrshire to play golf. The day was beautiful and the sun was shining, but the bitterly cold wind did not encourage me to play golf. Of course, I cannot entirely blame the gale because, in Calcutta, where I grew up, there were hardly any gales, but I had more enthusiasm for other sports than golfing. The golfing competition is always a social programme at medical conferences, and one in which I have never participated.

Gales and snow were common occurrences when l lived in Scotland, north England and the Midlands, and thus we became accustomed to winter snow and the associated lifestyle. Some Christians in Britain believe that a white Christmas brings good luck; the first white Christmas we spent in England was in Richmond, Yorkshire in 1969. My wife and I were the guests of a Rotarian, Francis Willies and his wife Lillian, and snow started to fall on Christmas morning when we arrived at their house. From their lounge, we saw the snow falling and soon the Yorkshire Dales were covered with snow. Francis told me that on a nice summer day the entire valley could be seen from their front garden and one could drive to the Lake District through the narrow, windy, tortuous roads of the Yorkshire Dales. This we did quite a few times in summer, spring, autumn and even in winter fog, and enjoyed the beauty of the nature all year round.

As members of the Rotary Club, we were involved in Rotary activities, especially during our stay in Rourkela, India. Thus we came across many people by involving ourselves in organising eye camps, visiting a leprosy colony and carrying out other social, welfare and international activities. If we talk about the Rotary Club of Rourkela, we always remember our close friendship with Rotarian Lall and his wife Sashi. Their premature deaths due to cancer shocked us very much. Mr Lall was always Santa Claus at the children's Christmas party at the Indo-German Club. Our children enjoyed the party and were very happy when he took out their presents from his bag.

Christmas seems to be more popular in England than in Scotland, but New Year is more popular in Scotland than in England. On Christmas Day and Boxing Day in England, most shops, offices and public transport used to be completely closed. This reminds me of Dr Sadhan Datta, a gynaecologist from India who was stranded at Heathrow Airport and rang everywhere for a transport to go to South Wales. Nowadays, this pattern is changing; there are supermarkets and public transport that stay open for limited hours and run limited services on Christmas Day and Boxing Day. It might be due to economic reasons and also the changing patterns of Britain's multicultural society. Whatever the reason, it certainly helps overseas visitors who are in this country for the first time.

Snowfall is not so common on the southern coast of England in comparison with Scotland, the north of England and the Midlands. This is something we have certainly noticed while living in Hampshire. Scientists are warning that if global warming goes on at the present rate, then the south coast of England will develop a Mediterranean climate.

Meteorological records began in Britain in the seventeenth century and 1684 was the probably the coldest year, when the River Thames in London froze. However, I have never seen a river completely frozen in Britain.

When I visited Moscow in winter 1971, I saw for the first time the frozen River Volga. The former USSR is a land with long rivers and most of them freeze during the major part of the winter, although there is low snowfall in much of the country.

The River Volga is the longest river in Europe, about 2,300 miles (3,700 kilometres) long. It originates in the north-west of Moscow and runs through the heart of Russia, falling into the Caspian Sea. In the winter, the river freezes to a depth of about six feet. I saw a few boats and small ships locked by ice in the Volga, and children and teenagers were playing or ice-skating on the frozen river. I was told that during the winter people commute across the river on foot to work or to shops in the city. However, the worst winter ever recorded and faced by Muscovites was in January 2006, when the temperature fell to – 30°C. To escape the harsh winter, many people took shelter underground in Moscow's metro stations. Metro stations in Moscow are spacious and beautiful; I was told that underground stations were built in such a way so that many people could take shelter during an air raid, if Moscow was bombarded, which happened during the second world war. The irony is that, since then, instead of air raids, harsh winter forced people to take shelter in Moscow's underground.

To visit Canada in winter is a great experience, and one cannot ignore the magnificent views of snow-covered forest, the Canadian Rockies, frozen lakes, rivers and waterfalls. Unfortunately, winter claims more than one hundred lives each year in Canada, due to winter storms and cold temperatures.

Blizzards are very common in some parts of Canada. The Saskatchewan blizzard of February 1947 was the worst winter storm that had ever occurred in Canada. It lasted for ten days and buried an entire train in a snowdrift one kilometre long and eight metres deep. The same year, there was also a serious snowfall across the UK, and snow fell every day from 22 January to 17 March 1947. There were several snowfalls of sixty centimetres or more, and the depth reached 150 centimetres in upper Teesdale and the Denbighshire Hills. There were snow five metres deep on blocked roads and railways; it is extremely rare to see a train completely buried in snow, but it happened in March 1891 at Dartmoor in England, and in January 1978 in northern Scotland.

When I was living in north-east England, Teesdale and Weirdale were my favourite spots that I often visited. The Teesdale and the Tees waterfall is where the River Tees originates and then passes through Barnard Castle, Darlington, Stockton on- Tees, Yarm and Middlesbrough, ending in the North Sea.

In 2004, when I revisited Weirdale and Teesdale, I found lots of changes. There were definitely improvements in the roads, and by the roadside there were snow stump poles which guided the driver in snow or blizzard conditions. Moreover, there were dual carriages, from Bishop Auckland to Penrith.

Some of these areas of north England and Scotland are transformed into ski resorts in winter with winter snowfall.

1962–63 was the coldest period of winter in Britain since 1740 and a minimum temperature of -22.2°C was recorded in January 1963 at Braemar, Scotland. A cold winter kills people. According to 1998–1999 statistics, the UK has the worst winter death rate in the developed world; 30 per cent in comparison to Norway and Sweden, which is nearer to 8 per cent. The elderly and young children are particularly vulnerable to cold-related illnesses. It is quite common for National Health Service (NHS) admission rates to be higher in winter – mostly elderly and vulnerable groups like children and people with chronic chest disorders – resulting in an overstressed NHS. Preventive measures such as flu immunization might help to reduce the mortality and morbidity of the vulnerable groups. However, fuel costs and poverty also hit the elderly and other vulnerable groups, who struggle to pay their fuel bills. To help these people, in February 2000 the British government introduced a winter fuel payments scheme. Although limited, this scheme will hopefully have some impact on the reduction of cold-related illnesses in winter. Time will tell.

After the last great freeze in 1963, the winter of 2009-2010 has been described as a next great freeze in Britain. Much of the UK was blanketed in heavy snow. The satellite picture showed that from John O'Groats to Lands End, the whole of the British land mass was covered with snow. Canals and lakes were frozen. By the 7 January 2010, it was reported that twenty-two people had died because of the freezing conditions. This episode started on 16 December 2009 as part of a wave of severe winter weather in Europe and it continued the months of January, February and even in March/April 2010 in some places.

The lowest temperature was -22.C, which was recorded in Attnaharra, in Scotland. The temperature was similar to the North Pole. The deepest snowfall was on the Pennines in northern England; it was fifty-seven centimetres (22inches) of snow. In southern England where we live, in January 2010, the snow fell to a depth of thirty centimetres locally.

It was the coldest winter for more than thirty years in England.

I remember when the heavy snowfall started on the 5 and 6 January and for two to three days, we were not able to go out because of continuous snow. We live on the top of a hilly area: no cars were able to drive up or down. Most of the neighbours kept their cars at the bottom of the hill on the main road, near the village hall. For nearly two weeks, there was no milk delivery at home and we were kept indoors with minimum shopping.

The community spirit was high and many of the neighbours came out with shovels to clear the road and also helped other people by clearing pathways and assisted in pushing or driving the cars. Most people, including me, went to the local shop by foot and it was also nice to see many people walking instead of driving. I met a neighbour whom I had never met before and also a neighbour with whom I was able to talk after two years. I also got a car lift from one of our neighbour for some groceries shopping from our local supermarket but while returning we had to leave the car at the end of the road as it was difficult to drive up the hill to our door.

There were some panics amongst shoppers in case bad weather prevented them from purchasing further supplies over the next few days or weeks. Part of the motorway near our house was also closed.

There was much disruption to transport, with the cancellation of trains, closed airports and roads, trapped passengers, closed schools, business failure, power failure, reduced attendances at businesses and offices. The number of people in shops was reduced, as were the cars on the road. Car accidents and falls by pedestrians on icy roads and pavements were increased, hence hospital admission rose.

Football pitches were fully covered with snow and so many games were cancelled, although many fans tried to help to clear the snow from the pitches. The demand for salt and sand on the road was heavy and so reserves were spread mostly on to the main roads and motorways. The majority of the side roads were deprived of grit, making them difficult or impossible to drive. Electrical failure due to the cold temperature was responsible for a number of passengers trapped on the Eurostar train in the Channel tunnel. With reduced numbers of cars on the road, there were less carbon emissions on the road but at home there was increased demand for gas and electricity resulting in greater carbon consumption.

In 2009-2010 winter storms were reported across the Northern Hemisphere. It was very cold and the snowy spell of weather resulted in falls of snow, hail and sleet in the form of blizzards, ice storms and snowstorms. Arctic winds and snows gripped China, across Russia to Western Europe and over USA and Canada. In the north of China the temperature fell to -32.C. The extreme cold and ice condition killed sixty people in India. Western Russia, the Baltic States and north-eastern Europe were affected by the deep freeze. Europe became a frozen continent. Forty centimetres of snow fell in the Netherlands, thirty centimetres in Germany and twenty to thirty centimetres in France. Icebreakers were deployed on the canal, harbour and inland seas of the Netherlands. German householders were advised to stock up on food, medicine and drinking water. The temperature in Poland was -22.C and winter deaths in Poland were up to 140, mostly due to hypothermia. In Norway the temperature fell to -42.C.

The day is short during winter in the polar and cold climate zones. In Scandinavian countries in winter, one can hardly see the sun or daylight and the prolonged darkness often affects their health. Psychiatric illnesses and the suicide rate are high in Sweden.

The greatest snowfall recorded over a year (1971–1972) was 31,102 millimetres (1,224.5 inches) and it happened in Paradise, Mount Rainer, USA. Snow and blizzards are not uncommon in the USA. American history tells us how the European settlers and pioneers faced and found difficulty in the harsh winter conditions, even though they were accustomed to snow in their homelands. The east coast and its cities are the American Snow Belt areas, and its residents have to negotiate severe winters quite often.

In America, I had severe snowfall experiences on the west coast rather than the east coast, when I visited Yosemite National Park in California and Lake Tahoe in Nevada–California in March 2006.

Yosemite is famous for its high concentration of falls in limited spaces and is full of lakes, ponds, rivers and alpine areas. The park area is 1,189 square miles (3,081 square kilometres) and most of the areas are 2,000 to 13,123 feet (600 to 4,000 metres) above sea level. There are seventeen falls, but I visited only one: Yosemite Falls, which is 2,425 feet (800 metres) high. I had a spectacular view of the falls in the midst of heavy snowfall and harsh weather, but I did not see any frozen rivers, ponds or falls.

The road conditions were so bad that, while coming back, our coach was guided by a park ranger and the tyres were chained. Most of the roads are usually closed in winter, that is, October to May, but visitors come all year round. High season for visitors is usually in summer, but winter visits are also popular because of winter sport activities.

Lake Tahoe is located 198 miles (316.8 kilometres) north of San Francisco in between the eastern border of California and the western border of Nevada. The day I arrived in Lake Tahoe, it welcomed me with heavy snow. Snowfall is not uncommon in March. The annual snowfall is 152 inches (308.08 centimetres), most of which occurs between December and March. Lake Tahoe never freezes because of its depth and constant water movement. This I noticed when we were cruising on the lake in a paddle steamer. The lake is twenty-two miles (35.2 kilometres) long and twelve miles (19.2 kilometres) wide. The deepest part is 1,645 feet (501.39 metres) and the colour is turquoise-blue; I could see quite far through the clear water of the unfrozen lake. The border line of the two states cuts through the lake. On the shore we stopped in a hotel where I had tea in front of a fireplace, standing with one foot in Nevada and the other one in California. I took an interesting photograph of the hotel's outdoor swimming pool, where I saw the well-marked dividing line between Nevada and California on the floor of swimming pool.

The Nevada portion of the swimming pool was more frozen than the California side, but no area was completely frozen.

December 2009 saw some of the worse snow and blizzard conditions in meteorological history that slams the East Coast in the USA. Philadelphia, Baltimore and Washington DC were hit the worst: in Washington DC, the winter snowfall was 139 centimetres (54.9inches), the snowiest on record for 110 years. Road conditions were so bad and hazardous that snow-plough operation were suspended and in some places could not be operated as the ploughs were broken. Airports were closed, trains were disrupted, buses were not running, and many roads were closed. Cars sat buried in the snow, power was cut off and schools, businesses and offices were closed.

From childhood, if anybody talks about snow, two things have attracted me: the snow house and the snowman. A snow house is the igloo, mostly used by the Eskimos, who live in the Arctic region, namely northern Canada, Alaska and Greenland. Other people who live in the Arctic region are the Chukchis of Siberia and the Lapps of northern Scandinavia. They all have to live a hard life by herding animals, hunting, fishing or trading. Since the discovery of valuable mines and oilfields, many outsiders live and work in the Arctic. The Arctic and Antarctic are the coldest areas of the earth, and very few people live in the Antarctic. The body's metabolism and nutrition are the important factors for survival in the Antarctic region.

Living in an igloo is much warmer than a tent, and how the Eskimos build the igloo used to surprise me. When I became aware of igloo-building technology, I discovered that to build an igloo is simple and is an interesting art. It needs a hard field of snow – hard enough to make solid snow blocks. The igloo is dome-shaped. Large blocks are used for the base of the dome and smaller ones on top. The edges of each snow block should be smooth and angled correctly to make a strong bond to the adjacent blocks, and a full circle of snow blocks is built with an entrance.

When I visited Greenland in August 2010, I found that Eskimos were no longer living in igloos and nowadays they live in houses and flats.

In Greenland the Eskimos prefer to be called 'Inuits', the indigenous people of the Arctic. It is said that they originated from Central Asia, migrating from the far east of Russia, then across Alaska and Canada, finally reaching Greenland about 4,500 years ago.

The Abominable Snowman of the Himalayas is called the Yeti, and supposedly lives in the mountains. There is a debate about the existence of the Yeti, as Western adventurers or researchers up to now have not been able to find any convincing evidence except some footprints, although local people believe they have encountered the Yeti in their daily lives. To the local people, the Yeti is an apelike creature and its existence was recorded in sixteenth century hand-written Tibetan manuscripts, kept at Buddhist monasteries in Bhutan.

I also heard that one of the high officials in the Bhutanese king's family kept in his private collection the remains of the body of a Yeti who was captured in tragic circumstances.

However, the search for a Yeti will go on until the human race is convinced about its existence, although its extinction cannot be ruled out, as the Himalayas are increasingly engulfed by human traffic with the growth of the tourist industry each year.

It is beautiful to see snowfall but it is hard to clear the snow afterwards. If one goes through the history of snow removal, one will find how people used to rely on the sun or mild weather to melt heavy snowfalls, which in some places is still happening. So is the use of manual shovels and salt. Then came the snowplough with horse-drawn carts, which was effective in clearing the streets in the eighteenth and nineteenth centuries. However, in the twentieth century, motorisation revolutionised snow removal and replaced the horse-drawn carts. The snow-removal fleets were modernised with trucks, ploughs, caterpillar tractors, blades, snow loaders, giant scoops and conveyor belts. The demand for salt and sand used to remove the ice left behind by snow ploughing is responsible for motorised salt-spreaders, which ultimately became the primary tool in fighting snowy roads.

The disadvantage of spreading salt is that it can cause corrosion to cars, roads and bridges. Road traffic accidents (RTAs) are not uncommon on icy roads; however, in hospital accident and emergency departments, I saw more casualties from having slipped or been injured on icy roads, footpaths or pathways rather than RTAs, but serious or fatal accidents were more due to RTAs. Although there are some disadvantages of spreading salt on roads, its role in preventing road accidents is great, and this I realised when my car skidded on an icy road in Yorkshire, though fortunately we were all safe.

There have been developments in advanced snow-removal technology in the aviation and railway industries to clear runways and rail tracks. Lastly – and most important of all – is the accurate forecasting of climate and weather conditions. Space and satellite technologies have revolutionised weather prediction. Fortunately, nowadays the public is made appropriately aware of impending hazardous situations by radio, television and other media.

Chapter Eight
Fog and Smog

Smog in a Chinese city

Fog reminds me of horror films, or movies showing how a prisoner or smugglers are escaping, or a ship lost in the midst of fog and mist. Fog also plays an important role in the writings of Charles Dickens and Sir Arthur Conan Doyle.

My worst experience of fog was in 1969, when I was crossing the Pennines in the north of England with my wife. We were travelling from Blackpool to Harrogate; this was before the motorway had been built over the Pennines. The road we took was through the Yorkshire National Park. The road passes through the dales, hills, forest and moors and, on a beautiful day, it is a beautiful drive. Unfortunately, on that winter evening it was not. As we approached the Pennines, the road became more narrow and winding and sometimes there were slopes or drops on both sides of the road. I was driving on the left-hand side of the road through dense fog. It was a dark winter night and, as we went higher and higher into the Pennines, the fog became denser and denser, and the traffic thinner and thinner. In some places, the visibility was virtually nil. This made it difficult to drive. Along the winding road, I was driving slowly, guided by catseyes, which were in the middle of the road, dividing the traffic. There was hardly any traffic. I felt the road would never end and this was my worst nightmare. However, the unique catseye road reflector saved us from any roadside disaster.

It was a Yorkshireman who invented the catseye, which I learned from another Yorkshireman, whom I met at a social gathering and to whom I was narrating my horror experience of driving through the dense fog. The catseye's inventor was Percy Shaw (1890–1976) and the idea that made him famous originated in an incident that took place not far from where I was driving my car on that day. In 1933, Shaw was driving to Bradford in dense fog and had a close escape when the beam of his headlights was reflected in the eyes of a cat sitting on a roadside fence. The eyes of the cat saved Percy Shaw, and his invention of the catseye road reflector device saved us from roadside mishap.

Accidents arising from poor visibility due to heavy fog while driving at a high speed are not uncommon. Speed control, fog detection and appropriate warning systems help to prevent such accidents. On motorways nowadays, it is possible to control vehicles with fog warning signs and limited traffic speeds, but on other roads it is still an issue. More focus is needed on how to prevent non-motorway road accidents due to fog, but in the past thirty years, the improvement of fog lights in the form of driving lights, street lights, parking lights and fog warning signs have no doubt made it safer and more comfortable to drive.

It is also beautiful to see when amber streetlights glare on the road in the foggy winter evening and cars pass through with their fog lights on.

Initially, I could not understand why in some places the streetlights were an amber colour. Later on, I found out that amber streetlights are fog lights which only light up the road under fog. It is most commonly a fifty-five-watt halogen lamp, and lenses and reflectors are important for the construction of the lights. However, this does not necessarily mean that fog lights are amber-coloured and driving lights are white. Fog lights are amber because of the colouring of the lens or the reflector. Unfortunately, colouring may reduce the intensity of the light. So, for this reason, white lights may be better fog lights than amber lights. However, the advantage of the amber lights is that they cause fewer glares off snow, which is useful in severe weather conditions (such as mist, fog and snow) where illumination of the road is essential. Fog and mist can be extremely dangerous, and are responsible for many accidents and fatalities. Visibility is the greatest problem.

Fog consists of cloud in contact with the Earth's surface and is therefore merely suspended water droplets whose formation depends upon cooling causing condensation. Fog and mist are the same, and the only difference is 'visibility'. Mist has a visibility of between 1,000 and 2,000 metres. Fog is defined as when the visibility on land is 200 metres or less, but at sea or for aircraft landing or take-off purposes; it is when visibility is 1,000 metres or less. However, the international definition of fog is when the horizontal visibility drops below one kilometre (0.62 miles), and heavy fog is when the visibility is not more than one quarter of a mile.

Ground fog is mostly radiation fog and sea fog is advection fog. Radiation fog occurs mostly on land on a clear night with light winds and does not last long after sunrise. It may occur any time between early autumn and late spring, but is unusual in summer. Advection fog is thicker, more widespread fog that lasts longer than radiation fog and often appears over the sea; it can occur throughout the year and at any time of the day.

The other forms of fogs might be of interest. They are as follows:

Upslope fog is hill fog that occurs on hillsides or mountain slopes and on mountaintops can cause freezing fogs. This fog can last for many days over a large area.

Steam fog is the most localised form of fog and is commonly seen around deep, large lakes and in polar regions. It occurs in late autumn and early winter and often causes freezing fog.

Valley fog is a kind of radiation fog that usually occurs in winter on mountain valleys and can last for several days.

Precipitation fog is also known as frontal fog, and forms as raindrops fall from warm air into cooler surroundings below, which then evaporate and cause fog.

Ice fog occurs when the droplets have frozen and the temperature falls below freezing. It is found near the Arctic or Antarctic regions with a kind of precipitation.

Fog exists on mountaintops, hills, slopes, valleys, the sea, lakes, rivers, swamps, tropics, glaciers and in polar regions, and is responsible for reducing visibility. Fog is more dangerous when it is localised because sometimes the driver is suddenly or surprisingly caught in the fog. Fogs can hit air and rail traffic, resulting in cancellation, diversion, delays and disruptions. Twice I have been caught in a fog that struck air traffic. Once, I was flying from Jersey to Teeside Airport via London but, due to fog, the evening flight from Heathrow to Teeside was cancelled. I was stranded at Heathrow and there was a big commotion there because of the delays, disruptions and cancellations. On the other occasion, I was travelling from Delhi to Sophia, the capital of Bulgaria. Due to heavy fog, the aircraft was not able to land at Sophia's airport. Instead, it landed at Verna airport, near the Black Sea resort.

San Francisco in the USA is famous for summer fog. During most of the summer, chilly Pacific air with a shroud of fog invades San Francisco Bay and the temperature hardly rises above 70°F. The fog and low clouds cause low visibility and local people are quite used to it, but summer is a bad time for visitors. The famous spots for fog are Candlestick Point and the Golden Gate, which I visited when I was in San Francisco in 1988 and 2006, but not in the summer. The Golden Gate strait is the most famous landmark in San Francisco and it is particularly famous for the suspension bridge.

This was built in the 1930s, and it was first time that this kind of bridge had been built, although cable suspension bridges have been around since the nineteenth century. It took four years to complete and opened in May 1937. It is 1.7 miles long, linking San Francisco with the mainland. The main suspension area is 4,200 feet long, which I walked through and of which I took some photographs.

On the Indian subcontinent, though the thunderstorm is a common phenomenon, the question remains: 'Does fog follow an outbreak of thunder?' This is probably rare, but an incident that occurred when I was a teenager in Calcutta supports the existence of 'thunderstorm fog'. This was in 1956 and I had gone to Calcutta airport to see my father off. As a journalist, he was flying to Helsinki, Finland, to attend a conference of the International Journalist Association. In those days, to go to the airport and to see the aircraft flying, landing and taking off used to interest me. There was no terrorist threat, no strict security restrictions. One could go fairly close to the aircraft to see it taking off. However, that night, the weather was bad and a thunderstorm was causing great problems, meaning that international flights leaving Calcutta were delayed. We returned home, leaving my father and his colleague at the airport. However, I was surprised to see my father the next morning at home. He had returned, as the flight had been cancelled because of the thunderstorm and fog. I was puzzled, as the thunderstorm was over, but the flight had still been cancelled; at that time, I had no knowledge of 'thunderstorm fog'. I had the impression that the thunderstorm and fog were two separate phenomena. However, now I know that dense and persistent fog, resulting from the very damp air after a thunderstorm ceases, may occur. Although rare, 'thunderstorm fog' is now a well-documented phenomenon.

Fog also reminds me of a photograph of Calcutta's morning fog, showing the Victoria Memorial Hall and the *Maidan,* indicating that winter had arrived in Calcutta. This used to appear every year on the front page of one of the renowned Calcutta- based daily newspapers. I am not sure whether this tradition is still ongoing or not, but I have heard that Kolkata experiences more smog than fog nowadays. The Victoria Memorial and the *Maidan* are both important landmarks in Kolkata, and have stood proudly since the days of the British Raj.

Smog is a mixture of fog and smoke. Smoke is visible vapour, which generates from burning particles such as cigarettes, coal and factory chimney emissions.

Smog is usually caused by urban air pollution and the sources are: road vehicles like cars, buses and lorries; non-road vehicles like planes, trains and ships; industries, business and households. The pollutants are mostly sulphur, nitrogen oxides, carbon monoxide and ozone. Smog is harmful to health and most big cities face this hazard. In nineteenth century Britain, industrialisation was responsible for fog and smog. London was the most commonly affected city. Historically, the worst air pollution disaster was the 'London Fog' of December 1952, whose details I did not know until I did an epidemiology and air pollution class while doing an MSc in Occupational Medicine at London University. The London Fog killed 4,000 people. Sulphurous pollution was responsible for smog formation. Soot, sulphur dioxide, chemical and petrol fumes covered the London sky, Londoners suffered from cardio-respiratory disorders and hospital admissions went up. The concentration of sulphur dioxide (SO2) reached nearly 4,000 mg/m3, ten times more than the maximum level set down by the World Health Organisation (WHO) as safe to breathe for one hour. The Clean Air Act of 1956 came into force. For the past thirty years, sulphurous pollution in Britain has been drastically reduced. This is possible because the Clean Air Act forced people to use gas, electricity and smokeless fuel in place of coal, and there has also been a reduction in industrial pollution. Moreover, the traditional heavy industries like steel, coal and shipbuilding no longer dominate British industry. This has resulted in a change in the industrial map of Britain.

However, according to the WHO, one in five of the world's population is exposed to unhealthy levels of sulphur dioxide and the Third World is the major victim. This reminds me of when I was involved in the environmental pollution control programme of the Rourkela Steel Plant in 1979. This is one of the steel plants under the Steel Authority of India and Rourkela is its steel city, situated in the state of Orissa, India. This was after I had finished my MSc in Occupational Medicine in 1976 at London University.

My interests in research lead me to occupational medicine and it was Professor Richard Schilling, the doyen of occupational medicine, who introduced me to this speciality, although I had other options. Professor Richard Schilling (1911–1997) was then Director of Occupational Health at the TUC Centenary Institute, London School of Hygiene and Tropical Medicine. I was in his last batch of students. He retired in 1976 and died of cancer on 30 September 1997.

After obtaining my MSc and DIH, I became an occupational physician and got a job in occupational medicine in the British Steel Corporation, Teeside. This was in the days before the Faculty of Occupational Medicine was formed. In April 1978, the Royal College of Physicians set up the Faculty of Occupational Medicine and, in 1984, the General Medical Council (GMC) in the UK agreed to recognise the FFOM and MFOM as additional qualifications for the purpose of registration, and I was awarded membership (MFOM) in that year. After a spell of working with the British Steel Corporation, I joined the Steel Authority of India (SAIL) in 1978. In the Rourkela Steel Plant, there was a question of sulphurous pollution; furthermore, not long before, the roof of one of the areas of the steel plant collapsed due to the accumulation of dust, triggered by bad weather. The analysis of air and environmental pollutants showed an increased level of sulphur dioxide, carbon monoxide and nitrogen oxides. Besides the Rourkela Steel Plant, the other steel plants in India also had environmental and modernisation issues, and a Boston-based US company took up environmental pollution measurements in the majority of the steel plants in India, under SAIL. I visited the company's laboratory in Boston in 1987 and saw some of the pollutants analysis. The steel plants underwent modernisation and I hope, as a result, the sulphurous pollution is now less troublesome in Rourkela, as well as in other steel plants and steel cities.

In 1988, although I left the Steel Authority of India, and came back to England, I was still travelling between India and England, carrying out some of my unfinished work. In January 1989, I was in Delhi and, one winter morning, I was travelling in the car along with one of the director generals of the Industrial Research Council of India. We were going to his office. The winter fog and smog was covering the Delhi roads and skyline.

I was sure that the emissions from road vehicles were responsible for such smog. We discussed the Delhi pollution level including ozone, and the director general talked to somebody over the phone about environmental measurements. In the evening, I heard on the radio that pollutants such as suspended particles, as well as the levels of sulphur dioxide and nitrogen oxides, were high. However, somebody told me that Delhi now has less vehicle emission pollution because of the introduction of strict vehicle emission laws. I am not yet fully convinced on this because, whenever I visit Delhi, Mumbai or Kolkata, I have found that air pollution is still a big issue. I thought that Kolkata was worse than Delhi or Mumbai.

I found travelling by foot on Kolkata's roads very troublesome. In some places footpaths were occupied by hawkers, vendors and food stalls. I faced congested roads with irregular traffic and crowds. Most of Kolkata's main roads were dug up to accommodate the metro, India's first underground railway system. The construction of multi-storeyed buildings and overhead road systems was making some of the areas worse. The transport system was a big problem: the buses were crowded and taxis sometimes difficult to find. Trams, which are pollution-free, were once the proudest and most luxurious transport system in Calcutta; they have now become the worst transport system, and there is talk of abolishing them.

I saw black smoke coming out through most of the vehicles' exhaust pipes. In addition, the road dust and a lack of civic sense were making things worse. The morning and afternoon road- watering I found to be carried out on very few roads in Kolkata. On one occasion, I felt that I was going to be choked. I took my handkerchief out to protect myself from poorly combustible exhaust fumes and road dust. This reminded me of a documentary film on Japanese cities; where people walk or ride bikes with face masks on to escape the pollution. Those living in Kolkata often complain of health problems like coughs, eye irritations, allergies, breathing problems, headaches, etc. Air pollution reports showed the presence of high levels of sulphur dioxide in winter and increased quantities of suspended particle matter and nitrogen oxides. Air pollution control is urgently needed in Kolkata. In 2003, I heard that Kolkata created the Anti-Pollution Cell, but I am not sure how effective it is in controlling air pollution.

Kolkata has changed considerably in its 300-year history. Once it was the capital of British India and now it is only the capital of West Bengal, a state in India. As part of a developing country, Kolkata is developing slowly, some say more slowly than other cities in India, which I found when I visited India in 2006. The city is still one of contrasts, of the old and new. Kolkata is densely populated, polluted, and in some places very dirty. Problems of waste disposal and refuse collection are still a nuisance, as roads are sometimes blocked with refuse, which is not very impressive to visitors. Kolkata's green belt is reducing and ponds and paddy fields are disappearing, as the city is extended towards the east, south and north, once the suburbs of the original Kolkata. The salt lake area has been developed to accommodate Kolkata's growing population. The metro railway system is completed and is running efficiently from Tollygunge to Dum Dum, and also extended further. Major roads have been connected with overhead roadways. There is a great increase in the middle class population. More cars are on the roads. Pollution and population are still the major issues, and more emphasis must be given as to how Kolkata can be kept green and clean.

Urban air pollution is no doubt a major threat to human health. In Asia, urbanisation is the main cause for deterioration in urban air quality in twelve major cities in Asia. Those are Bangkok, Beijing, Kolkata, Delhi, Karachi, Manila, Mumbai, Osaka, Seoul, Shanghai, Tianjin and Tokyo. High population density, a high number of motor vehicles and intense industrial activity have a great impact on air pollution on those cities. The study on cities' environment pollution showed some pollutants exceeded the standards laid down by the WHO. High levels of suspended particles, sulphur dioxide and nitrogen oxides were detected, and the worst-affected cities were Beijing, Tianjin, Kolkata, Delhi, Tokyo and Shanghai.

According to the Guinness Book of Records, the most polluted major cities of the world are Mexico City, Beijing, Xi'an and New Delhi. Mexico City is top of the list and its acute levels of sulphur dioxide, carbon monoxide and ozone exceeded the WHO's standard, with a moderate to heavy pollution of lead and nitrogen oxides.

When we visited China in 2004, I saw the effects of air pollution on Xi'an , Beijing and Shanghai. It appeared to me that the nature of Xi'an's air pollution was different from the type and nature of air pollutants in Beijing and Shanghai cities. Xi'an's air pollution might be more due to emissions from coal power plants and factories.

Xi'an is one of the cities in China where the old city walls still remain in the midst of a modern city. The walls form a rectangular area. On each side, there is a gateway and each gateway has three towers. At each of the four corners is a watchtower. The centre of the town is the bell tower. We had a VIP welcome at the city gate, in the old traditional way.

Xi'an is famous for Emperor Qin's Mausoleum and the Terracotta Warriors Museum. This well-known place is about twenty miles east of the city, which we visited when we stayed at Xi'an. The Terracotta Army was discovered in March 1974 by local farmers drilling water well to the east of Mount Lishan. I bought a book on Qin's Terracotta Army from one of the surviving farmers who discovered the site accidentally. He signed his autograph on the book.

The Terracotta Museum is located east of Emperor Qin's Mausoleum. The museum consists of three pits with about 8,000 life-sized terracotta figures of warriors, archers and horsemen.

Some of them are in squatting positions and some are standing, holding bows, spears, daggers, axes and other long shaft weapons. Warrior figures were built in such a way that their faces and expressions differ from each other and all are dressed according to their respective ranks of warrior.

We saw all three pits: the first is the largest one and contains 6,000 figures of the Emperor's main army with infantrymen, chariots and horses. The second consists of 1,400 figures of cavalry and infantry along with chariots. The third contains a command unit of sixty-eight figures of officers and a war chariot drawn by horses.

Emperor Qin's Mausoleum is situated at the northern edge of Mount Lishan, which is 1,256 metres high and covered with lush green trees.

Emperor Qin Shi Huang, China's first Emperor, built his empire after unifying the various warrior clans and the country. After many great achievements, he wanted to be immortal. He was obsessed with the fear of death. He used to take pills that contained mercury, assuming they would promote longevity and even immortality, but they were in fact harmful to his health. In search of the elixir of immortality, he even went out to sea, as he was told that in the middle of the sea there were three supernatural mountains where immortals lay. However, in spite of various attempts, his search for immortality was in vain. He died in 210 BC at the age of fifty, and his dead body was buried alongside a great amount of treasure and artefacts in a tomb with a scale replica of the universe, complete with gem-encrusted ceilings representing the cosmos and flaming mercury representing the great earthly bodies of water. The Yellow River, Yangtze River and other rivers of the country were produced with quicksilver. Pearls were also placed on the ceiling of the tomb to represent the stars and planets. The tomb at present has not been opened up or excavated, and most of the findings are the result of satellite investigations on pits and tombs. Scientists also conducted investigations on mercury levels at the site and found high levels in the soil of Mount Lishan. Mercury contamination might be one of the reasons why the tomb has not yet been unearthed or excavated.

However, in contrast to Xi'an, Beijing's and Shanghai's air pollutants were mostly from petrochemical smog. I also tried to find out why those cities have such high levels of air pollution, especially the petrochemical smog.

With the economic development of China, bicycles in the main cities are being replaced by cars. I noticed the cities' skylines are covered by great tall buildings, and Beijing is booming and bustling with skyscrapers and a sophisticated wide network of roads. Shanghai has monumental tower-block buildings. Each evening, these buildings and the whole city of Shanghai are lit up and the city at night is one of the main attractions for tourists. For the generation of electricity, the city still relies on coal power plants.

The cities of China are already overpopulated and their population is increasing, with migrants arriving from the surrounding areas every day. Shanghai is one of the most densely populated cities of the world, having a population of more than seven and a half million; more than three million being the migrant population. China's manufacturing industries are booming and the principal exports are machinery and electronic goods, garments, footwear and toys. No wonder China has problems with air pollution. To overcome the effects of pollution on health, some of the cyclists I noticed wore masks. To me, the cities' skies looked dull and smoky even at 10 or 11 a.m. I was told that two years ago it was worse.

However, pollution was an important issue in 2008 Olympic Games, in China as it was held in Beijing. To improve its air quality, China banned the millions of cars from the roads as part of one of their measures prior to Olympic game.

The Western world and some of the cities of Asia and other developing countries suffer from petrochemical smog, which is also known as summer smog.

Petrochemical smog or summer smog is caused by the chemical reaction between vehicle emissions and sunlight, which irritates the eyes, nose and throat. This type of smog is a mixture of hydrocarbons, nitrogen oxides, sulphur dioxide and carbon monoxide emitted from road vehicles, industry and power stations. Ozone levels increase on sunny days and this affects the health by causing irritation, mainly to the respiratory tract.

Lead can also cause air pollution, affects the health and is usually present as an anti-knock agent in petrol. Some of the countries of the world have already eliminated lead content from petrol.

To tackle carbon emissions from cars, Brazil has looked for an environmentally friendly renewable source of fuel, and that is ethanol. I noticed this when I visited Rio de Janeiro in August 2007. Brazil is now the largest producer of ethanol, which the country makes by fermenting and distilling the sugar cane crop and then using the liquid for car fuel. Seven out of ten new cars sold in Brazil are now accommodating the 'Flexi-Fuel' system, so that individuals can fill the car with either ethanol or petrol. All petrol pumps are selling pure ethanol or gasohol. Gasohol contains a blend of petrol and 25 per cent ethanol. Ethanol is no doubt a cleaner-burning fuel than petrol. It causes less pollution and smog, and less maintenance is needed for cars, as ethanol is kinder to the engine. Besides sugar cane, ethanol can also be produced from corn, sugar beet, wood chips, grass and organic waste. Like Brazil, China,

India and Mexico must look for such an alternative renewable fuel in order to prevent further air or transport pollution.

Unless we control and reduce road traffic and its emissions, the harmful effects of smog on our health will continue to rise, alongside global warming. More emphasis also has to be given to alternative energy sources like solar power, wind turbines and tidal power instead of the traditional coal power station. Some countries are already planning to explore alternative sources of energy from outer space, such as tapping energy from the moon. This might be feasible if the appropriate technology and money is available. However we do not know what impact there will be on the universe and its environment!

Chapter Nine

Lightning

Lightning

At the age of thirteen or fourteen, when I was first becoming interested in the art of photography, somebody told me that I would win an award in photography if I could catch the lightning phenomenon with my camera. I then tried several times to catch the most spectacular view of lightning, the majestic flashes that light up the sky, with my Rochiflex camera, but I failed. It was because the lightning strikes and roars as fast as it disappears, and it puzzled me for several years how quick a lightning strike travels. My frustrated mind used to ask me, 'Is there a way by which one could extend the lightning travelling phenomenon for a few more seconds, so that I could capture a good picture of lightning that flashes across the sky?'

Since ancient times, lightning has fascinated as well as scared people.

The ancient Greeks used to identify lightning as the weapon of the father god, Zeus, the most powerful Greek god. The god's missiles always struck high buildings and trees, and Xerxes, the ruler of ancient Persia, made this interesting observation.

When lightning strikes, it reminds me of an incident in my childhood. It was Sunday afternoon, I was in my maternal uncle's house and suddenly I heard thunder roar. I looked through the veranda and saw lightning strike a coconut tree and hit a middle- aged woman who was standing underneath it. She fell. This happened so suddenly and quickly that nobody realised what was happening. Within a minute, a crowd surrounded her. I could not see whether she was injured or dead. However, I saw an ambulance, which came and took her to a nearby hospital. Later on, I heard that she had died.

Every year, thousands of people are killed by lightning throughout the world. The next morning when I went to my school in Calcutta, it was a great topic for discussion. I told the teacher in front of the whole class about the incident of the lightning that I had witnessed on the previous day. There was a discussion about what would happen if lightning struck our school building. The teacher assured us that we were safe, as our school building was protected by lightning conductors. A device connected our school building with the other nearby great building, the Calcutta Palace of the Maharaja of Tripura, which would divert the lightning discharge away. At that time, I was probably too young to understand the negative and positive charge of the electricity and details of the conducting pathway.

This was probably the result of Benjamin Franklin's famous kite experiment in 1752. Benjamin Franklin (1706–90) erected the first lightning conductor, which was his own invention, on his house in Philadelphia in 1752. However, the device was denounced by many, especially by some religious leaders. The exact mechanism of the lightning phenomenon is still mysterious. Since Benjamin Franklin's time, scientists have understood that electrical charges can slowly accumulate in clouds and then create brilliant flashes when the stored energy suddenly discharges. Franklin's conductor rod device either channels the discharge or diverts the lightning away. This device does not prevent the lightning strike. The scientific community is still not able to control lightning and its damage.

Franklin as a scientist is internationally known, but I was not sure that Benjamin Franklin was the same person who played a significant role as a statesman in the American Revolution until I visited Philadelphia in 1997. This was partly due to my ignorance, particularly my poor knowledge of American history. In June 1997, I went to New York to attend a medical conference, held at the Rockefeller University in New York City. At the end of the conference, I went to Crossville, Philadelphia, to spend the weekend with our old family friends Ram and Usha Iynger. Ram was an ex-director general of the Indian Scientific Research Council, based at New Delhi, and is now living in America.

The fourth day of July is Independence Day in America and, on that day, they took me to the historical site at Philadelphia where the American Declaration of Independence took place and the American constitution was written. Benjamin Franklin was one of the authors, along with Thomas Jefferson and John Adams.

The history of human civilisation has witnessed many scientists and entertainers become politicians or statesmen, but it is uncommon that the politician becomes a scientist, a film star or an entertainer. The most recent examples of the former group include Professor Abdul Kalam, the eminent Indian scientist, who became the president of India, and the famous Hollywood star Ronald Regan (1911–2004), who became the president of the USA and held the office for two terms. The popular American president died at the age of ninety-three. Perhaps popularity more often tempts and leads to political power.

Philadelphia is the largest city in the state of Pennsylvania, situated in the industrial belt of the USA. It was summer, so we drove to the coastal area, where I saw a nuclear power plant. For electricity generation, the

USA is very much dependent upon nuclear and hydroelectric power besides coal power plants.

In 1986, when I was in the USA, I visited the great Niagara Falls, where the Niagara hydroelectric power plant is situated. The power plant supplies power to the eastern part of the USA and southern Canada. Niagara Falls consists of three massive waterfalls on the Niagara River in the eastern USA, on the border between the USA and Canada. The Niagara River is only thirty- five miles (fifty-nine kilometres) long. It originates in Lake Erie, ends in Lake Ontario and is 12,000 years old. The three falls on the river are the Horseshoe or Canadian Falls (height: 173 feet/53 metres; width: 2,200 feet/670 metres), the American Falls and Bridlevela Falls. The combined height of the American and Bridlevela Falls is 182 feet (56 metres) and the width is 1,100 feet (326 metres). It seems that Niagara Falls is very wide but not so high, as 168,000 cubic metres (six million cubic feet) of powerful water falls down in a cascade. Crossing the falls from the American to the Canadian side by boat and observing its beauty was a great experience for me. Every year, millions of people visit this natural wonder of the world, as the place is famous for its unusual beauty and hydroelectric power.

On 15 August 2003, there was a fire in the Con Edison power plant on the American side of Niagara Falls, resulting in a blackout that paralysed the cities of New York, Cleveland, Detroit, Toronto and Ottawa. Fifty million people were affected. The Canadian authorities were of the opinion that the fire in the Niagara plant was caused by a lightning strike, which was completely denied by the American authorities. Not long ago, in June 2002, lightning was responsible for a serious malfunction at the nuclear power plant in Florida, USA, although the plant was safe as it was designed to take lightning strikes. The same year, a power plant in Kosovo was hit by lightning, resulting in a plant explosion and injuring several workers.

When I was in India, power failure used to be a chronic problem in certain places. It was due to outdated plants and equipment, and large consumer demand. For power, India is usually dependant on traditional coal and hydroelectric power. The situation used to get worse if a thunderstorm hit the power system. Lightning hits towers, the structure of transmitters, power lines and plants.

I also worked for a utility company, named Scottish Power, which generates electricity and supplies it to central and southern Scotland, the north-west of England and north Wales. From time to time, there are power failures resulting from thunderstorms and lightning that hit the transmitters or power lines. I have made several enquiries regarding the lightning faults on the Scottish Power network as I have seen some of their workers engaged in repair work, as a result of lightning. I was informed by one of the engineers from Scottish Power that there were 381 faults per year due to lightning and 133,507 customers were interrupted per year by voltage levels, although the recoveries were quick. The records over five years showed only one power failure, lasting for six hours. This was the longest one. However, these are not as common as in some parts of the world, namely the USA, India, Africa, Australia and South America.

I tried to find out the reason behind this, and looked at the frequency distribution of lightning strikes across the earth. In Scotland and parts of England and Wales, lightning strikes 0.1 to 0.5 times per square kilometre per year, but in other parts of the world, which are in the thunderstorm zones of the world, it ranges from 8 to 50+ times per square kilometre per year.

The world's highest thunderstorm area is Kampala, Uganda, in Africa, which has thunderstorms on average 290 days per year and is said to be the most electrified area of the world.

Whatever the mode of power generation, power failures are not uncommon in the thunderstorm areas of the world, with heavy costs for the electricity companies. It usually costs the USA alone $1 billion annually.

It is very difficult to give exact figures on how many times individual lightning flashes hit the ground and to what extent they destroy life and property throughout the world. According to some sources, there are about 1,800 thunderstorms and approximately one hundred lightning bolts strike the Earth every second at any given time. About 10,000 people are killed and 100,000 people are injured each year by lightning. The human body is a good electrical conductor. When electric current passes through the body, it causes convulsions (violent shaking of the body). The victims are sometimes thrown into the air with a dreadful crash landing, leading to multiple injuries. The skin is burned and muscle is scorched. The heart stops if the current passes through the heart. Breathing comes to a halt, and the victim loses consciousness if the current passes through the brain. The long-term effect on lightning strike survivors' health includes phobia, loss of memory and concentration and personality changes, although it needs more research.

Lightning causes forest fires. Lightning starts about 10,000 forest fires every year in the USA and, in 2002, Russia reported seventy- seven lightning-induced forest fires in the far east of Russia. Forest fires due to lightning are also not uncommon in Europe, Australia, Canada, South and Central America. Fires threaten timber, habitat and human settlements and, according to some scientists, fire is also responsible for sun shields, which might cause secondary pollution.

Forest fire smoke can cause eye irritation, sore throat, cold and cough symptoms, chest pain, shortness of breath, wheezing and airway congestion. It might cause acute worsening of asthma, bronchitis and carbon monoxide poisoning. The USA is famous for redwood trees, which are the tallest trees in the world and can survive for more than 1,000 years. However, most of the redwood forests in America have long been destroyed due to forest fire and heavy logging. The only surviving redwood forest found in America is the Muir Woods National Park, which I visited in March 2006 and is only forty-five minutes' drive from San Francisco. I saw tall redwood trees, some of them exceeding 250 feet. These towering trees are still standing in the midst of the damp forest, despite the past history of frequent logging and forest fire. As they survive for many years, their trunks are very big and they easily reach ten to fifteen feet in diameter. Their bark is so thick that they sometimes survive forest fires, because it is slow to burn, and I noticed quite a few half-burned trunks in the middle of the national park.

I also saw a trunk that was more than 1,000 years old. The age of a tree is now easily identified by carbon-dating. Before carbon- dating, electricity had been used to measure plant growth. It reminds me of the Indian scientist, Sir Jagadis Chandra Bose (1858–1937), who invented an electrical device called a crescograph for measuring or recording plant movements and growth.

While I was in America, I saw a news item in the local newspaper that said that twelve miners in a coal mine in West Virginia had died after an underground explosion. The company believed that it was triggered by a lightning strike. The lightning probably ignited methane gas that had accumulated in an abandoned part of the mine.

Lightning strikes aircraft and also disrupts their navigational devices. According to some estimates, on average, lightning hits a commercial airline once a year. However, most aircraft do not fly into lightning storms or fly through areas where lightning is present. This I have experienced twice: once, when I was stranded at Glasgow Airport, as my flight to Southampton was delayed by three hours due to thunderstorms and lightning in southern England. Cities such as Plymouth were lit up by multiple strikes. A team of fifty scientists using lasers, radar and balloons gathered to seek clues about the development of destructive thunderstorms. Luckily, nobody was injured except one child, but there were several reports of houses struck by lightning. I heard that a newly built hotel in Titchfield, Hampshire, caught fire when lightning struck it. Customers in southern England faced power failure a couple of times, but it was restored quickly. In the UK, the distribution of thunderstorms suggests that they occur more frequently in southern England than in the rest of the country.

On another occasion, I was flying from Newcastle to Glasgow. Suddenly, halfway there, our aircraft was caught in a thunderstorm and I could hear the pilot announcing, 'Be seated and fasten your seatbelts as we are in a thunderstorm.' Nobody was panicking, but the way the aircraft was swinging and flying, there was a certain feeling that our aircraft was lost. However, we landed safely and credit goes to the pilot. At the end, the pilot disclosed the truth that we were lucky to have been saved. To overcome the disaster, the pilot had to take the aircraft up quickly to high altitude and fly above the thunderstorm zone or above the storm clouds, in which he thankfully succeeded.

It seems that lightning presents a potential hazard to aircraft. Many people think that aircraft should be designed in such a way as to divert the current away from risk areas.

There are even incidents where lightning strikes the launch pad or near a space shuttle before launch.

Lightning occurs usually in two ways: one is within a cloud and the other between the clouds and the ground. The lightning that occurs within the cloud is called sheet lightning, and lightning that occurs between the cloud and the ground is called fork lightning.

Traditionally, it was thought that the lightning phenomenon was mostly restricted to the lower atmosphere. This concept has changed and it is now realised that the electrical discharges take place in the air up to ninety kilometres above thunderclouds. This high-altitude phenomenon has been witnessed by many from aircraft, from the ground and from space shuttles. Lightning has been seen in the atmospheres of Venus, Earth, Jupiter and Saturn.

Thunderstorms often occur at the end of hot, humid summer days. During this time, the warm moist air rises quickly and forms large, tall, dark clouds. Inside the dark clouds, the air current creates a strong draught and as a result, water droplets and ice particles rub against each other. Static electricity builds up as they bang into each other. Ice and water, which are heavier and negatively charged, accumulate at the base. The lighter particles that produce positive charges gather at the top of the cloud. The ground below the cloud is also positively charged. The electricity flows between the charges and the electricity is released as a flash of lightning, when the flows are neutralised. This depends on the degrees of difference of flow between the two charges. As the lightning strikes, the surrounding air is heated and the heat causes the surrounding air to expand, which explodes, resulting in a loud crash. The crash is the thunder and the flash is from the lightning. That is the reason that it is said that thunder rolls and lighting flashes.

The question is, does lightning occur before thunder? No, lightning and thunder occur at the same time, although we see the lightning before we hear the thunder. This is because light travels faster than the sound. Now I understand why as a teenager I was not able to photograph the spectacular view of lightning and thunderstorms.

Thunderstorms scare toddlers and children, and it is not uncommon that some of us at that age wake up in the middle of the night with fear. We are able to overcome the phobia and go back to sleep after hearing our favourite stories and being comforted by our parents or granny. We certainly had such experiences with our children, especially with my youngest daughter, Tanya. Tanya was a toddler when she was in Rourkela, India, where, after the summer heat, thunderstorms were a common occurrence. I remember in the middle of the night when the thunder roared, she invariably climbed up out of her cot bed and came into our room to take shelter in between us in our bed.

The safest place during lightning storms is inside a building or a car with the windows rolled up. Unsafe places are open spaces and fields, standing under trees, water-related activities, using heavy road equipment, telephone booths, and so on.

Recently, a lightning strike injury has been reported while a person was using a mobile phone. It raises the question of whether mobile phones increase the risk of being struck by lightning, and this needs further research.

However, two questions about lightning still puzzle me.

The first one is, is there an individual susceptibility to being struck by lightning? The second question is: as we know that there are some sorts of electricity always present in our body, what happens when a person produces excessive electricity or generates a high-voltage electricity current? Is one dangerous to him/herself or others? What happens to him/her when lightning strikes?'

As far as I could recall, most probably there was a person identified with high-voltage body electrical phenomenon, but I was not sure and so I did some research. I found in the Guinness Book of Records that there was a man called Roy Sullivan, an ex- park ranger from Virginia, USA, who was struck seven times by lightning, indoors as well as outdoors, until he died in September 1983. It was recorded that in 1942 he lost his big toenail; in July 1969, he lost his eyebrows; in July 1970, he seared his left shoulder; in April 1972, his hair was set on fire; in August 1973, his hair was set on fire again and he seared his legs; in June 1976, his ankle was injured; and, in June 1977, his stomach and chest were burnt while fishing.

Researchers and scientists are still trying to solve the puzzle of lightning. I was rather excited when I saw an article on lightning in May 2005 in the New Scientist journal. The article concluded that 'lightning comes from outer space and the X-ray emissions detected from lightning bolts confirmed this' and 'researchers are also convinced that lightning is caused by some kind of runway breakdown and this could be triggered by cosmic rays.'

Lightning was once considered to be a divine power, but scientists are nowadays able to understand lightning more and raise the hope of controlling it. This could be achieved by early and accurate weather prediction, a better global lightning detection network, better protection devices (lightning arrestors or grounded shielding) and, lastly, efforts in controlling the lightning with lasers.

Chapter Ten

The Sun, the Solar System and Solar Eclipses

Astronomical Observatory, Hawaii

Among all the natural phenomenon, light is the most important and the sun is the most useful source of light and heat. The sun lies at the centre of the solar system. Blocking of the light of the sun or the moon causes an eclipse. The word 'eclipse' is a Greek astronomical word which means 'failing' and it can be total or partial. When Earth blocks the sun's light from the moon, it is called a 'lunar' eclipse and when the moon blocks the sun's light from the portion of the earth, then it is called a 'solar' eclipse.

A 'total' solar eclipse can occur when the moon is in orbit between the sun and the earth, and the sun cannot be seen until the moon has moved across the sun's disk. The disk of the moon appears black and is surrounded by the sun's corona.

An eclipse of the sun always occurs during the day, but not every eclipse of the sun is a total eclipse. This is because sometimes the moon is too small to cover the entire sun's disk; the moon's orbit around the earth, which is oval or elliptical in shape, plays an important role and, as the moon orbits the planet, the distances varies. When the moon is on the near side of its orbit, it appears larger than the sun and, if an eclipse happens at that time, it will be a total eclipse. On the other hand, if the moon is on the far side of its orbit and the eclipse occurs at that time, it then appears to be smaller than the sun and cannot completely cover the sun. A ring of sun remains around the moon. This type of ring-shaped eclipse is called an 'annular' eclipse. A 'partial' eclipse means that only part of the sun is covered by the moon. Lastly, the 'hybrid' eclipse is where the moon is at such a distance that the eclipse is annular in some areas and total in other areas.

Since the creation of the solar system, the sun has provided heat and light to the earth, and has a huge magnetic field. The surface of the sun has three components:

a) Photosphere – this is the visible surface of the sun and its temperature is approximately 6,000°C.

b) Chromosphere – this is the pink ring just above the photosphere and its temperature is about 10,000°C.

c) Corona – this is the surrounding area with temperatures of 1–2,000,000°C.

The period between 1949 and 1956 were interesting years for me, as I was becoming more inquisitive about the sun and solar eclipses. I was told how Sir Isaac Newton's (1642–1727) own experiments on colour risked damage to his eyes by viewing the sun with the naked eye. Later on, he discovered the optical system for analysing the sun's spectrum, a great scientific achievement. It is well known that permanent eye damage or blindness can occur after viewing the sun with the naked eye, especially when looking at annular or partial phases of total eclipses without using an appropriate filter. It is definitely dangerous to the eye.

It is the infrared radiation that damages the retina of the eye, and this should be avoided during solar eclipses.

Chronic exposure to the sun can cause photo-dermatitis and also increase incidents of skin cancer among Caucasians. Ultraviolet solar radiation is sometimes responsible for photo- dermatitis or photo-aggravated dermatitis, which I saw in India. I also noticed this among the Uros Indians in Peru, when I visited the floating reed islands at Lake Titicaca. The dazzling high-altitude sunlight is responsible for the ultraviolet radiation. Cusco, the ancient capital of the sun-worshipping Incan Empire, is found to have the highest ultraviolet light level in the world today. So it was not surprising for me to see such solar radiation-related skin hazards. Lake Titicaca belongs to the governments of Peru and Bolivia, and the lake is the highest navigable lake in the world. To get there, we travelled by road from Cusco for 380 kilometres, climbing 4,313 metres (14,108 feet) and then descending until we reached the shores of Lake Titicaca. The lake is at 3,812 metres (12,507 feet) above sea level. The lake is shallower in Peru, and deeper in Bolivia. The reeds usually grow in shallow water and so the artificial islands made of reeds are located on the Peruvian side. There are forty to forty-three such small islands in total. About three years ago, one of the islands was destroyed completely by fire, and seven people unfortunately died. The fire was an accident, the result of a lit candle. As there is no electricity on these islands, the inhabitants rely on kerosene, candles, etc. However, since the fire the islanders have stopped using candles.

The Uros Indians are some of the most ancient tribal people in South America. They rely on fishing, hunting, taxidermy and tourism for their day-to-day sustenance. We visited one of the reeds islands, where we were welcomed in the traditional way. We saw their traditional way of life, which is very much dependent on reeds. Their huts, boats, furniture, watchtowers, fuels, arts and crafts are all made from reeds. We also had the opportunity to observe them making some of these items.

During 1949 to 1956 – my school days in Calcutta – I tried to observe various eclipses. On the occasion of solar eclipses, especially during a total solar eclipse, people used to stay indoors, behind shut doors and windows. The children were told not to use the toilet to pass water or to move the bowels during eclipse time. This was because of superstitions rather than science.

However, my scientific curiosity never allowed me to stay indoors, but I was warned not to look at these great astronomical events directly with the naked eye. In those days, eye protectors made of good solar filter glasses were not easily available. A home telescope was out of the question. We had to rely on water or other reflective surfaces, and even tried to look at the eclipse indirectly by creating a hole in a piece of pitch board.

The longest solar eclipse that has occurred in the twentieth century was on 20 June 1955, lasting seven minutes, eight seconds. I remember I tried to observe that solar eclipse from Calcutta, but I could not, as the centre line passed through Sri Lanka, the Indian Ocean, the Andaman Islands and south-east Asia (where the total solar eclipse was visible), and ultimately disappeared into the Pacific Ocean.

When I went to university to study medicine in 1961, I had to forget my interest in astronomical science, although I could not ignore the advancement of space science, which has always attracted me. This continued whether I was in Britain or India, until the occasion of the total solar eclipse on 16 February 1980. At that time, I was working in India but visiting England. After spending a couple of months in England, I was returning to India and I stopped at Rome. There I met a group of scientists, mostly from the USA and Europe, who were on their way to the Orissa coast in India to observe the double solar eclipse from the site where the great sun temple, Konarak, is located.

Most of them were carrying telescopes as well as other photographic instruments. I was tempted to accompany them. However, I could not as I already had work commitments at the Rourkela Steel Plant, which is fairly near to Konarak, Orissa, India.

'Konarak' means 'Place of Sun' and the temple is dedicated to the sun god, Surya. It is located on the shores of the Bay of Bengal. Bathed in the rays of the rising sun, the chariot of the sun god has twenty-four wheels, and the sun god rides and moves across the heaven with seven horses pulling the chariot. There are three images of the sun god, positioned to catch the rays of the sun at dawn, noon and sunset. The base, walls and roof of the temple are carved with images, including some in an erotic style.

Sailors used to call the Konarak sun temple the 'Black Pagoda' because it was supposed to draw ships into the shore and cause shipwrecks. Some sort of magnet and its magnetic field was most probably responsible for the shipwrecks. The Black Pagoda was used as a navigation landmark by mariners sailing to Calcutta. The main temple collapsed and the main idol of the sun god on which the sun rays fell in the morning is said to have been removed by some Portuguese sailors.

Built in AD 1250, the Konarak temple is now protected under the World Heritage list of UNESCO. To save it from deterioration, an initial attempt at restoration was made in AD 1903, by the then-British Lieutenant Governor of Bengal, by filling up the temple interior with stones and bricks. After India's independence, further excavations and restoration took place.

I remember that, in 1949, I spent one month of the summer holiday at Puri, Orissa, with my parents, grandmothers and my younger brother and sister. This was an important trip for me, as it was my first visit to a sea resort. As a child, to visit a sea resort, play on the seashore and swim in the sea was a great attraction and experience. The Konarak temple is thirty-four kilometres from Puri, and my father was planning to visit Konarak. I was adamant that I would accompany him. The party was supposed to leave our guesthouse at 4 a.m. The next day, I went to bed early, hoping that my father would take me on his next morning trip. When I got up in the morning, I found that he had left without me or any other family member. I was so disappointed that I cried for nearly a whole day. At that time, I did not realise that it was not a safe environment in which to travel, especially for children and women. Although the distance is short, in those days it was very difficult to travel to Konarak from Puri. There were no proper roads and the only way to travel was by bullock cart through the jungle. Along with my father on that morning, a team from the Indian film division, the Government of India Publicity Division, also went to Konarak to make a documentary film, which we saw later in the cinema halls of Calcutta, focusing on the excavation and restoration work. In those days, there was no television.

However, this small incident made me more determined to visit the Konarak temple in later years, which I did in the 1960s, 1980s and the 1990s.

In the 1960s, I was a medical student, and at that time I visited with one of my old school friends, Ajit Mukherjee, who was working in Orissa as an engineer. The road was not too bad, although some portions of it were not tarred but made of red sand, brick and stone.

In the 1980s, it was completely different; the road to Konarak was fully tarred and easy to drive on. This time our children were able to accompany us, something that I could not do as a child. Moreover, a couple of friends from England also visited Konarak with us. They were Dr Michael Saunders and Dr Peter Newman; both were consultant neurologists from north-east England. When we were living in Middlesbrough, Michael and his wife Irene and their four children were very close to us. Irene was a consultant psychiatrist who was working in a local psychiatric hospital. They both spent some time in India, working in Vellore, southern India. One time, I worked in the neurology department under him at Middlesbrough and we also did some research work together when I was working at the British Steel Corporation and published a paper. In India, I travelled a couple of times with him. Michael is now retired and has become the Reverend Michael Saunders. This time, after attending a medical conference, Michael and Peter came to visit us at Rourkela. They were interested in visiting the temples of Orissa. From Rourkela, we went to Bhubaneswar, Puri and Konarak by car. The drive was bumpy, and on one occasion the car broke down.

My last visit was in the 1990s, and each time I visited I always noticed some improvements in the roads, as well as the temple. This time, I felt that, as a World Heritage Site, it had been made more commercial and more of a tourist attraction.

Lots of people have asked me, 'Why do you go to Konarak so often? Are you a sun-worshipper?'

My answer is that although there are various legends and myths behind the Konarak temple building, I was puzzled and searching for the right answer to some questions:

Why the Konarak temple was built on that isolated site, and was there any archaeo-astronomical explanation for it?

Was this the place where the maximum amount of sunshine was noticed by the medieval local people?

Was this the best site where the first morning sun rising from the Bay of Bengal is visible on the eastern horizon?

Was it a site where total eclipses were frequently visible?

Is it aligned with solar and lunar astronomical events, and was it used to predict solar eclipses?

Why was the temple responsible for shipwrecks? Was it the temple stones or was there any specific magnetic iron pillar amalgamated during the temple construction that might be responsible for the shipwrecks?

Did the stone alignment attract and reflect the sun's magnetic power?

Does it symbolise the sun's magnetic field?

Were the Konarak temple builder and astronomer aware that the sun had a huge magnetic field?

Did they build the temple bearing this in mind?

Since ancient times, there has been a leper colony near the Konarak sun temple, which still exists on the same site. Is there any connection between the sun, medicine and astronomy?

Modern scientists have discovered that, every eleven years, the sun's twisted magnetic field releases huge amounts of energy in the form of flares and prominences. Sunspots also increase during that time, affecting radio and satellite communications on Earth.

The ancient peoples always feared darkness and, when total eclipses of the sun occur, the sky darkens, the air cools, birds and animals react and the landscape changes with unusual colours.

The local people react on the basis of events, culture and belief. This unusual phenomenon most probably provoked a sense of fear and alarm among the local people; this leads to worship of the sun god and goddess and resulted in building the sun temple at that site. Since ancient times, many people have believed that the sun is a healer of many diseases (namely leprosy, blindness and skin diseases) and therefore sun worship was not uncommon in ancient civilisations. Eclipses were always feared by primitive man as a sign that a heavenly monster was engulfing the sun, and they were signs of the god's anger, resulting in diseases and disasters.

Some of the sun gods/goddesses of ancient civilisations are:

Utu for the Sumerian, Ra for the Egyptian, Surya for the Hindu, Ten suns for the Chinese, Lugh for the Celtic, Helios for the Greek, Sol for the Roman, Inti for the Incas and Ah kin for the Mayan and Amaterasu for the Japanese.

In 1980, at Rourkela, I did not notice much impact of the solar eclipse and also I was not sure how successful the scientists had been on their scientific expenditure at Konarak on the solar eclipse. However, I gather that superstitious fears were provoked in some cities in India, like Chennai and Hyderabad, and resulted in public transport staying off the roads and people staying indoors until the eclipse was over. In the eclipse of 1980, there were criticisms of the media's role in India in creating false fear and alarm by providing superstitious or pseudo-scientific information rather than providing appropriate scientific information on eclipses and safe methods of watching them.

Fifteen years later, the picture was different; the media was successful in providing scientific information on the total eclipse of 1995, instead of creating confusion amongst the public with counter-information by astrologers or pseudo-scientists. Although I was in England at that time, I was told that a large number of people gathered in the streets of Calcutta to watch the phenomenon, which was close to totality, and many people travelled to Diamond Harbour, forty-eight miles from Calcutta, which was in the path of totality. The scientific community was happy and gained satisfaction, and the television was able to cover all stages of the totality, which was also observed by a large number of television viewers.

A total eclipse of the sun was seen in the same year over the temples of Angkor Wat in Cambodia. Angkor is a vast complex of stone temple ruins, sacred to both the Hindu and Buddhist religions, and other monuments. Embedded in the jungle of Cambodia, the site is spread over 400 square miles. This was once the capital of the great Khmer empire that spread as far as present- day Thailand, Laos and Vietnam; the civilisation survived there from the ninth to the fifteen centuries. However, I gather that some of the statues have since been stolen. The complex needs extensive restoration work, although rapid transition is ongoing because of increased tourist influx. In October 1995, nearly 6,000 spectators throughout the world saw the path of totality pass right over the Angkor temple complex, when the sun disappeared from sight and darkness prevailed for a few minutes.

Diamond Harbour, not far from the place where the Ganges meets the Bay of Bengal, used to be a tourist or picnic spot; I can remember that, at Diamond Harbour in the 1960s, I took a beautiful picture of the sunset on the Ganges that was well appreciated. My last visit to Diamond Harbour was in 1976, along with Michael Saunders, and it was an interesting visit, including a country boat ride on the Ganges.

I remember there was a difference in opinion regarding how to treat the beggars. Beggars surrounded us and demanded money, especially on seeing foreigners, which I did not like and did not encourage, although Michael was in favour of giving money. Beggars and street begging are not uncommon in India, and one will find them wherever one visits: in cities, temples and historical or tourist places of interest. Most of them are professional beggars. India must look into the rehabilitation of this kind of professional beggars, although Indian philosophy and culture is unlikely to support the banning of the beggars completely, which we see in many parts of the world.

However, I saw a glimpse of some kind of rehabilitation programme, when we visited Mother Theresa's places in Sealdah and Kalighat in the mid-1980s, along with Michel Saunders and Peter Newman. We saw a blind old woman lying in her shelter in Kalighat, who had been taken from the streets of Calcutta. Her family had probably abandoned her and she had no other choice except begging in the streets of Calcutta. She was given shelter, food and care, but her habit of begging for money with the typical posture (the right hand up) was still noticeable.

Mother Theresa was born in Macedonia on 27 August 1910, and at the age of eighteen joined the Sisters of Loreto and was sent to India in 1931. From 1931 to 1948, she was a teacher in a convent school in Calcutta. She left the convent school and decided to devote herself to work among the poor. She started her own order, The Missionaries of Charity, which takes care of the destitute, abandoned and dying people of all castes and religions, and used to give shelter to the poor, orphans and dying people picked up from the streets of Calcutta. She established an orphanage. I first came across her orphanage in 1965, when she took care of an abandoned baby from the casualty department of the medical college opposite her Mission of Charity establishment in Sealdah. Her work among the poor in Calcutta won her the Nobel Peace Prize in 1979. She died in Calcutta on 5 September 1997.

In India, girls are more likely to be neglected or abandoned than boys. Population growth and control of the population is a major issue in India, but it is more worrying that, over the last decade, India's male/female ratio has been increasing. In addition, sex-selective abortions, using ultrasonography, by some doctors and the killing of unborn female children illegally in some cities in India is making the issue worse. The girl/boy ratio will soon be 750 to 1,000 in some states in India. If the ratio continues declining in this way, there will be an ecological and biological disaster due to a shortage of women.

In India, *ghats* along the River Ganges are famous for their history, and it is interesting to visit the Ganges *ghats* at dawn or in the early morning, when one can see the devotees of the Sun God taking a bath in the Ganges and worshipping the sun. Hindus offering *pindas* (offerings to the dead) are also common sights.

My photographs of the sunrise on the River Nile, the midday sun at Stonehenge and the sun setting on the Ganges allowed me to think what the important issues for the ancient civilisations were; some civilisations gave importance to the sunrise, some to the sunset and some to midday. In the magnificent settings of the stones in the temples of ancient civilisations, the sunrays would fall and produce shadows on the wall that would usually predict the time of the day. This was seen in the sun temples of India and also found in other ancient civilisations. In the

Incan city of Machu Picchu in Peru, the Incan priest-astronomers used to predict the winter solstice and the return of the lengthening summer days by observing the shadow of a post cast on the surrounding walls by the last rays of the setting sun, and this was possible because of the way the stones were set.

Nowadays, many people use the Inca trail to visit Machu Picchu, but we travelled by train from Cusco. It took more than four hours to reach Machu Picchu, and we covered 110 kilometres of narrow gauge track in a diesel-powered engine. Our railway journey began with a zigzagging, steep climb. I have been to many places in the world where I have had to climb mountains on narrow gauge trains. However, most of them gradually wind their way up the mountain and take a long time to climb. I had never been on a track that climbs back and forth on a switchback system. It was slow, taking half an hour to climb out of Cusco. The train then passed through narrow valleys, mountain passes and tunnels. The snow-capped mountains, rivers and villages appeared and disappeared as the train rattled down the fertile plain of the Urubamba Valley, past Ollantaytambo and on to our destination, Aguas Calientes. From here, we had to take a twenty- minute bus ride to the lost city of Machu Picchu.

In the Incan city, we saw temples, palaces, dwellings, plazas, streets, paths, stairways and drainage channels. The city is divided into two sectors: agricultural and urban. My interest in archaeo-astronomy prompted me to explore certain areas, such as the Temple of the Sun and the observatory, the hitching post of the sun (the Intihuatana), the Temple of the Three Windows and finally the Mortars Room. The 'Temple of the Sun' is a semicircular tower, of which the lower part contains the royal tomb. There are two windows: one that faces east and another that faces south-east. Each window has four protuberances, each one in a corner. The sun shines directly through the eastern window on the winter solstice (21 June), and on the summer solstice (21 December) it shines through the south-east window. The 'Intihuatana' was mainly an astronomical observatory tool, by which the Incas observed sun movement. They used it to calculate the hours of the day and months of the year. The three windows in the 'Temple of the Three Windows' were guides to sunrise. In the 'Mortars Room', there are two circular stones which contained water for use as mirrors. The Incas may have used these stones for their astral observations.

Stonehenge, the megalithic ruin at a site near the town of Amesbury, Wiltshire, in southern England, has had a significant impact on archaeo-astronomical science in Britain. Stonehenge (3000–600 BC) is the only prehistoric stone circle found in the world with horizontal lintels across the top of the stones; the way the stone circle is made with shaped stone, lintels and jointing demonstrates perfect knowledge of geometry. However, up until now there has been no explanation found as to why Stonehenge was built on that site. Many believe that it was constructed as a temple of the sun and that, on midsummer's morning, the sun rose directly over the Heel Stone. It must have been built according to astronomical knowledge, and could have been used to predict events such as eclipses.

The last total solar eclipse of the millennium was on 11 August 1999, and its pathway was from the north Atlantic to the Bay of Bengal. The first place the eclipse touched land was in Cornwall, England, and the pathway was sixty miles away from Stonehenge, but the rain and cloud spoiled the visibility of the total eclipse. Since 1927, this was the first total eclipse that was supposed to be visible from the British Isles, and the area from where it would have been seen was between Plymouth and Land's End. Millions of people gathered for a glimpse of the spectacular event, which taxed the ecological system and local services through traffic jams, medical emergencies, lack of accommodation and disposal of piles of rubbish.

The excitement was also apparent in other parts of the country, and this I certainly noticed among certain groups of people at the Portsmouth hospitals where I was working at that time. On the day, I looked at the sky, which was clear, and the weather was not too bad. We did not expect to see a total eclipse, but I certainly noticed some changes in the sky and the sun. Our hospital photographer was busy taking some photographs and no doubt he managed to shoot some spectacular views through his camera.

After leaving Britain, the eclipse passed through the countries of Europe, namely France, Belgium, Luxembourg, Germany, Austria,

Slovenia, Hungary and Romania. Then it crossed the Black Sea and passed through Asian countries like Turkey, Jordan, Syria, Iraq, Iran, Pakistan and India before disappearing in the Bay of Bengal. Many people in Europe enjoyed at least part of the eclipse, but unfortunately the monsoon clouds spoiled the chances of seeing the eclipse in Pakistan and India.

The total eclipse of the sun on 29 May 1919 was important for the great scientist Albert Einstein (1879–1955). The remarkable photographs of the eclipse confirmed Einstein's general theory of relativity. With regard to Einstein, I always remember the great Indian scientist Satyendra Bose (1894–1974), whose contribution to the quantum world was a landmark in physics: the 'Bose– Einstein' statistics. I met Bose in the mid-1960s at our house at Calcutta.

The importance of astrophysics, astrochemistry and solar physics are growing, but my interest in archaeo-astronomy and astromedicine is not faltering. It means more visits to the ancient sites of the world. So, in March 2008, I went to Mauna Kea, the sacred mountain of Hawaii, which some Hawaiians believe to be the home of their gods and ancestors. Ancient Hawaiians were great navigators and they were among the pioneers in the ancient world who used sun, moon and stars to guide them between islands in the vast Pacific. They aligned the position of the rising or setting sun with marks on their canoe's railings.

Many historians believe that the ancient Hawaiian astronomers observed the skies from the top of Mauna Kea by using alignments with prominent features of the mountain. They also observed the tide, winds and tsunamis from the mountain top. Modern-day astronomers also consider Mauna Kea to be the best site on Earth to study the universe. This is because of the height, shape and location of the mountain. Moreover, due to the dark sky, dry air, stable atmosphere and the lack of cloud cover, it forms the ideal place for conducting detailed studies. Mauna Kea stands at a height of 13,796 ft (4,205 m) and is located at about 20° north latitude in the middle of the Pacific Ocean. This means all of the northern sky and much of the southern sky is visible from the summit. Given its height, the brightness of the stars is not affected by the degradation that is normally caused by the atmosphere.

The history of modern astronomy at Mauna Kea began in 1967 when a telescope was first constructed at the summit. This was thanks to the University of Hawaii, whose 2.2 m telescope was funded the National Aeronautics and Space Administrations (NASA). Since then, many countries and organisations have established their own telescopes and astronomical observatories and these have resulted in a number of discoveries on the solar system. Visiting such sites was like a pilgrimage to me. I travelled to the summit in a four-wheel drive as part of a group on 29 March 2008. Mauna Kea summit lies in between Hilo and Kona and one has to travel from one of these two locations. We took Highway 11 to Kona and from there it was Highway 190 which joined up with Saddle Road which took us to Mauna Kea.

The Visitor Information Station was six miles up the road at the height of 9,200 ft (2,804 m) and this was where we stopped for an hour before climbing to the summit. The stopover allowed us to acclimatise to the high altitude and reduce the risk of altitude sickness. The road from the Visitor Information Station to summit was winding, steep and eight miles long. The first four miles were unpaved but after that it was a little easier as the road was well paved. We reached the summit as the sun was setting. It was a cold evening with clear skies and pure fresh air, and we were surrounded by the snow-covered summit. We were above 40% of Earth's atmosphere and the atmospheric pressure was approximately 60%. This meant that there was less oxygen entering the lungs and with moderate exertion, there was a high likelihood of altitude sickness occurring. Fortunately, none of our group reported shortness of breath or any other symptoms.

At the summit, we tried to identify the various telescopes which belonged to different countries and organisations and they are shown in Table 4.

As I am getting old, I am becoming more interested in science rather than medicine which was my passion in my school days. I am reminded of my admission interview at a medical college in Calcutta in 1961. During the interview, the chairman of the interview board asked me, 'What is your first name?' I replied 'Asim [Ashim]'. He then said 'What is the meaning of Ashim?' and I answered 'It means boundless (the object which has got no limits or boundaries).' He then asked me to

Table 4: List of Mauna Kea Telescopes and Observatories

Year	Countries/Organisations	Type
1968	University of Hawaii (UH), Hilo	0.6 m educational telescope
1970	Institute for Astronomy (UH)	2.2 m telescope
1979	NASA	3.0 m infrared telescope
1979	Canada, France, UH	3.6 m telescope
1979	United Kingdom	3.8 m infrared telescope
1987	California Institute of Technology (Caltech)	10.4 m telescope within the Caltech Submillimeter Observatory
1987	UK, Canada, Netherlands	15m Maxwell Telescope
1992	Caltech/California University	10m telescope within the Keck Observatory
1993	National Radio Astronomy Observatory (NRAO)	25m Very Long Baseline Array radio
1996	Caltech/California University	10m telescope within the Keck Observatory
1999	Japan	8.3m Subaru telescope
1999	USA, UK, Canada, Argentina, Australia, Brazil, Chile,	8.1m Gemini Northern telescope
2002	Smithsonian Astrophysical Observatory, USA/Taiwan	8 x 6m Submillimeter Array

give an example and as I was deciding between ocean and universe, the chairman replied, 'It is the universe.'

I am not sure whether the chairman was suggesting that I took up space science as a career path rather then medicine, but this did not alter my choice at the time and I was given a place in a medical college. There is no doubt that great advancements are taking place in space science now, with opportunities to explore and see more, including possible life in space. Such advancements are only viable because of the continuing advances in technology across the world.

The Sun is the centre of the solar system and nine planets (Mercury, Venus, Earth, Mars, Jupiter, Saturn, Uranus, Neptune and Pluto) with their moons orbit the Sun. Mercury is the nearest planet to the Sun and Pluto is the furthest, smallest and coolest planet. Jupiter is the biggest planet and the Venus is the hottest. Since ancient times, astronomers have observed the sky, initially with the naked eyes, then by telescope and now the space probes and Robots which explore the planets. Space travel by man started when Russian pilot Yari Gagrin orbited the Earth on 21 April 1961. The American Neil Armstrong was the first person to step onto the Moon on 20 July 1969. The first manned space station was Salyut 1 launched in 1971 by the then USSR and in 1981 the USA sent the first space shuttle Columbia in the orbit.

Rockets are important for space travel and without such powerful devices it is not possible to lunch satellites, probes and astronauts into space. Once the Rocket lifts off, the rocket boosters fall off and then the fuel tank falls away. A rocket has to reach 40,000 kilometres per hour (kph) to escape from Earth's gravity. Once it is out in space, the speed of the rocket drops to around 29,000 kph to stay in orbit. At the end of the mission the shuttle drops its speed and re-enter into the earth's atmosphere. Prior to landing the shuttle turns off its engines glides onto the selected runway.

Space shuttles like Discovery, Atlantis and Endeavour have made many successful missions into space but the missions of the Challenger and Columbia shuttles ended in disaster. The shuttle Challenger exploded shortly after lift-off in 1986 killing all seven astronauts on board; the shuttle Columbia disintegrated along with its seven crews in 2003 while re-entering the Earth's atmosphere. It was Columbia's twenty-eight mission when the accident happened. It is sad that USA's first shuttle which was active since 1981 lost like this. It seems that space exploration can not take place without tragedies. As of 2007, the statistics show that nineteen astronauts have been killed in flight accidents, eleven astronauts died in training and at least seventy-one ground personnel were killed during launch-pad accident. Astronauts wear space suits as they are exposed to various hazards in the extreme conditions of the space and also to enable them to adopt themselves to the Earth's environment once they are returned. They can suffer from decompression sickness, fatigue, balance issues, disorientation, sleep disorder, psychological disorder, loss of muscle mass, loss of bone density, immune system disorder and so on. To control such hazards or to prevent ill health among astronauts, the importance of space medicine is no doubt crucial.

Chapter Eleven
Droughts, Famine and Starvation

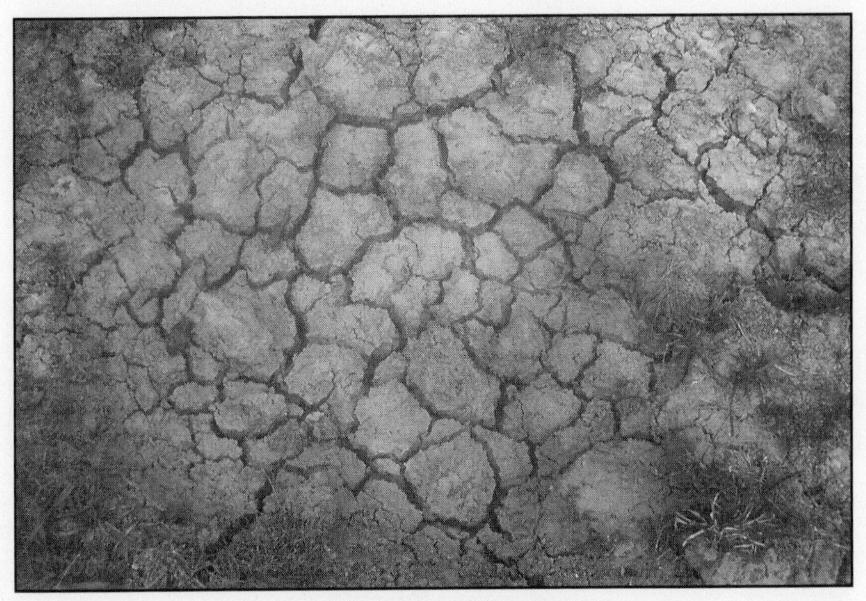

Drought-affected farmland

I was born during the Second World War, when the Bengal famine killed three million people in 1943. My mother used to narrate the story of how the people from the countryside and surrounding villages came and begged in the streets of Calcutta and died of hunger.

In front of our house, malnourished children, women and men used to shout: '*Ma, Ma vikhha dey na ma*' ('Mother, mother, please give us some alms'); '*Katho din khai, ni ray ma*' ('We are starved, we have not eaten for weeks'); and '*Phen dey na ma*' ('Please gives us some rice water').

Rice is the staple food of Bengal, and the normal procedure of cooking rice is to boil it and, when it is cooked, the boiled water is drained or thrown away. It seems that people were so starved that, for survival, they were begging for the rice water that was usually thrown away. However, there was no shortage of food but there were increased numbers of black marketeers and grain holders.

Although food prices were up, there was no lack of food in the cities and the people working on the railways and other essential services were able to feed themselves and their families because of the introduction of rations or supplies of certain food and essential items to the employees.

The presence of the military in Bengal in the Second World War made some people in Calcutta rich quickly. While the city was booming, the countryside was starving. The ordinary people in the countryside could not understand why there were shortages of food in spite of good crops. The people used to say, 'All food has gone to the Burma front, as there is a war with Japan'.

I vaguely remember that some people survived starvation but were killed by diseases such as typhoid and cholera. I thought that it was one of the worst man-made famines of the last century.

With the booming of the cities, some people became rich so quickly that their lifestyle was changed completely and they used to throw away surplus food into the drains. Some of their riches were short-lived. When the war was over, one family we knew became so poor that they were living on the poverty line and were hardly able to maintain day-to-day living. During my childhood, my mother used to give the example of this family and the Bengal famine if we wasted any food while eating – so we had to eat the food on the plate whether we liked it or not, and had to finish it.

I also gave similar advice to our children when they were young – saying that many people in the world did not have a square meal a day, especially in some of the poorest countries of the world, where hunger, poverty and starvation are not uncommon – so that they should not waste any food. On the other hand, while there is no more hunger, wastage of food is common in the Western world and other rich countries of the world. Instead of malnutrition, over-nutrition is the problem now. Obesity is crippling Western society. I remember in 1970, as a junior doctor, I was working in a hospital in Blackpool, England, and once I was dining with a healthcare professional who said, 'There were more healthy people in Britain during wartime because of wartime food rationing.' Whether this statement was true or not, certainly for the past thirty-six years I have seen increased numbers of obese young people in Britain, and this difference is invariably due to eating habits, the type or nature of food and the lifestyle of present-day British people.

The most famous work on the 1943 Bengal famine was done by the Indian economist Professor Amartya Sen. He worked on deprivation theory and demonstrated that millions of people could die from hunger when there was no shortage of food in the area. He was the Master of Trinity College, Cambridge, UK and won the Nobel Prize in 1998.

Since ancient times, it has been well known that war, natural disaster, drought and famine are linked with hunger and starvation. Some of the epidemics of certain diseases (namely typhus, cholera, typhoid and dysentery) are closely associated with famine, and people who survive starvation might die from disease.

Cannibalism has also been reported following famine.

The earliest recorded famine occurred in ancient Egypt and the Middle East. The Roman Empire also did not escape from famine and, in 436 BC, Rome itself was affected. Migration was not uncommon and cannibalism, death and starvation wiped out certain civilisations in South and Central America, and in the Pacific regions, although the Inca civilisation was famous for its extensive food storage system to overcome the effects of famine.

Since the seventeenth century, India and china have been notorious for food shortages, overpopulation and famine. Recorded famine in India goes back to the eleventh century and it continues into the twentieth century. The last major famine disaster in China was in 1959-61, when many millions of people are said to have died.

In this century, I was surprised recently to hear that bumper crops in the Punjab, India are going to rotten or eaten by rats because of the inadequate storage facilities in the country. Punjab is the state in India where the green revolution has been extremely successful and every year there are surplus crops. The irony is that, on the one hand there are many parts of the world including India where food prices are going up and millions of people are malnourished, and yet on the other hand, there is no place to store surplus grain. This reminds me my visit to *Golghar* in Patna in Bihar, India which is a granary constructed in 1786 to store surplus grains against famines, and is rarely used and even when it is used, it is barely filled.

It seems that to prevent the wastage of such bumper crops, India should give priority to building more storage, including the restoration of the old granaries.

In medieval Europe, famine was a common occurrence. Britain was affected ninety-five times and the French witnessed seventy- five or more famines. War and famine affected Russia in the eighteenth and nineteenth centuries. In 1845–1849, the potato famine in Ireland killed millions of people and caused further millions to migrate to the USA. Drought, dust bowls and severe health hazards were common across the Great Plains of the USA in the 1930s.

Drought is still one of the major disasters responsible for famine, hunger and starvation in certain parts of the world, especially some parts of Africa. Drought means a prolonged absence of water and results from the shortage of rainfall in a particular area. It ruins harvests, leaving people without food and water. If there is no food supply, the area or the country suffers from famine.

According to the Oxford English Dictionary, 'famine' is defined as the 'extreme scarcity of food in a region' and, in the Encyclopaedia Britannica, it is defined as, 'Extreme and protracted shortage of food, causing widespread and persistent hunger, emaciation of the affected population and a substantial increase in death rate.'

In the twenty-first century, one might be surprised to still see news headlines such as:

'Nearly 30 million Africans could be facing famine within months.'

'Millions across Africa are dependent on food aid.'

Africa's famine countries are mostly Ethiopia, Eritrea, Mauritania, Angola, Zambia, Zimbabwe, Mozambique, Malawi, Lesotho and Swaziland.

Even though food aid is available through charity, sometimes it does not reach the right place or the right people. This is because of other factors, such as corruption and the mismanagement of food supplies, armed conflict, trade policies and environmental degradation.

In spring 2006, my eldest daughter Rumella went to Ethiopia to cover the drought for the BBC. Ethiopia is one of the countries in Africa most affected by famine. She wrote to me a brief travelogue of her journey to one of the drought-stricken areas of Ethiopia.

> We arrived in Ethiopia at the start of their rainy season, except it had not rained very much. The capital, Addis Ababa, the third highest in the world, after La Paz (Bolivia) and Quito (Ecuador), is located in the rich central highlands. As we drove to the hotel from the airport, I got a distinct impression of an important African capital with tall buildings and wide roads. Many of the buildings were reminiscent of the Soviet era, betraying the Marxist followings of the nation's leaders. Addis is the headquarters of the African Union and therefore has an impressive diplomatic presence as well. It was hard to tell, from the busy, bustling city, that there was a drought affecting the country. We had to drive 650 kilometres south to find it. As we travelled downhill from the central highlands, the land became more and more parched, with less vegetation. Gone were the lush farmlands and the cool lakes. All you could see for miles on end were acacia trees and termite hills. We drove into a southern town called Yabelo, where the highest ranking government official said that he had never seen such a drought in all the fifty years of his life. He said that the people living in the area were pastoralists and depended solely on their cattle for survival, but as there was no longer grazing land available, the cattle were dying. As we travelled further south towards the border with Kenya, we saw fields of cattle carcasses. They began with the odd carcass lying by the side of the road with vultures pecking at its innards. By the time we reached the border

> town of Moyale, they were spread around in dozens. We were 750 kilometres south of the capital, and it was a very different Ethiopia down here. Speaking to the people, one had the impression that they had accepted their hardships as their fate in life. As a result of the knock-on effect of the cattle dying (that is, families losing their main source of income), even one square meal was hard to come by. The UN's World Food Programme, Oxfam and Save the Children were among the few agencies located in the area to try and alleviate the problem by feeding people and looking for more long-term solutions. But with more than 360,000 affected in southern Ethiopia alone, it seemed hard for them to cope. Widespread malnutrition meant that babies and children were beginning to die. Back in Addis, when we spoke to the Prime Minister Meles Zenawi, he said that the deaths were unfortunate, but the drought was an annual problem and that the government was handling it fine. This is only one of the problems that Ethiopia seems to be facing; a rapidly increasing population and gradual land erosion has meant that a nation that could once feed itself cannot anymore. And the war (border dispute) with neighbouring Eritrea has been a serious drain on resources over the years. As I left the country, I could not help feeling that this land of Solomon and Sheba will have to do a lot more to relive its glorious past.

Ethiopia is one of the poorest countries of the world and I am proud that Rumella has been there, to see how the drought and famine affected the country. When she returned, she also told me that the day she was returning, she went to see a family after hearing that an eight-month-old baby was dead because of starvation; the family could not afford to feed the baby. The mother was not able to produce any breast milk as she was malnourished. Rumella was so moved that she gave £80 to the family from her own pocket before she left the place, and she showed me the photograph of the family and their hut. Some of her stories have been broadcast on the BBC World Service.

A rich country like the USA also suffers from drought and there was big news in the summer of 2002, with headlines like:

'US bread basket is drying up.'

'Nearly half of US suffers drought.'

'Senate votes to aid drought farmers.'

Some of the farmers commented: 'I have never seen it this dry before in my lifetime', and 'This is the worst drought I have seen'.

There was not a drop of rain in some states of the USA, and some of the ranchers were feeding expensive hay to skinny cattle to keep them alive. The groundwater was disappearing and the rivers and reservoirs were drying up. The dry, extreme conditions might have been responsible for wildfires starting, and at least five states declared full-scale drought emergencies.

In the USA, the main danger of drought is bush fire and not famine. Similarly in Australia, bush fires are common after prolonged periods of drought and high summer temperatures, destroying forests and homes, and injuring and killing people.

On the 7 February 2009 the worst wildfire in Australian history, killed at least 200 people in the state of Victoria injuring more then hundred people. Thousands of people were left homeless as hundreds of houses were destroyed by fire.

In Australia, the drought in 2000 was the worst seen in a century and this caused Australian economic growth to slow down. According to the experts, 10 per cent of Australia's farming land is now unusable as a result of drought. It was reported that harsh conditions were forcing some farmers to sell out. Many people blame climate change as being responsible for the drought.

Summer 2010 saw the highest summer temperatures in Russia since records began 130 years ago, with the temperature reaching more than 35.C (some claim nearly 40.C). The nationwide heatwave was responsible for drought and forest fires and it was Russia's worst drought in more than an hundred years. The extreme temperatures destroyed nearly a third of Russia's grain crops with the most affected areas located in central and European Russia, where almost 25 million acres of crops were destroyed. The wild fires engulfed more than 64,000 acres of forest and more than 2,000 homes and about 750,000 hectares (2,900 square miles) of land were affected or destroyed. The forest fires also caused abnormally high temperatures along with a high level of air pollution. 54 people died throughout Russia and Moscow was covered with a thick layer of smog. Some days the carbon monoxide levels exceeded 6.6 times the acceptable limits. Muscovites experienced breathing difficulties, stinging eyes and severe depression. The unusual climate conditions were responsible for high mortality rates in the city. According the city's health department, during the spell about 700 people died each day in contrast with the normal figure of 360 to 380 people. Muscovites were advised to keep the windows closed and to stay at home. Pregnant women, children and elderly people were advised to leave the city or to travel to the less polluted areas. More than 2,000 people were drowned as many tried to cool off in river, lakes and reservoirs round the country. The heatwave might have cost 5,000 lives in Russia.

In the present day, Britain does not experience true drought and famine, although there have been shortages of water in rivers and reservoirs in recent summers, resulting in restrictions on the use of garden water. Scientists blame global warming for these shortages.

Drought is still not uncommon in India and specifically I remember two places, Purulia and Kalahandi (districts in eastern India). I was a final-year medical student and, as I was waiting for the final medical examination, there was a drought in Purulia and famine in certain parts of Bihar. There was a team of doctors from the IMA who were going to that part of the country to do some medical relief work. I discussed with the IMA my intention to go there, and it was decided that I would join the team once I had finished my final examination.

Unfortunately, I was not able to join the medical relief team because the date of my exam was postponed. This was due to some of the students at Calcutta University demanding postponement of the final medical examination; they *gheraoed* (blocked the way) the vice-chancellor of Calcutta University. I felt that this student action was illogical, although this type of *gherao* was not uncommon in the name of student movements in those days. Of course, I did not support this and some of those who supported the postponement of the medical examinations did not realise how important the time factor and its impact on one's career could be. It certainly affected me, and I was unhappy because I could not join the medical relief team. It also delayed my coming to Britain. I was offered a pre-registration post in Scotland, provided I passed the final examination and fulfilled the other criteria. I was worried about my future, and so were some other final-year students. So we went to meet certain senate members of Calcutta University, pleading against the postponement of the examination. However, we were not able to stop it and the examination was ultimately postponed.

As far as the medical relief work at Purulia was concerned, I did not regret it too much, although I would have had some medical experiences, dealing with the victims of the drought- affected areas. The team who went there came back and reported that the medical-related problem in the drought area was not a big issue or a problem, and so there was no need to send any more medical relief teams. Moreover, some of the medical relief workers sustained injuries when the Land Rover carrying the team was involved in a road traffic accident.

This was about forty-four years ago and I understand that drought still occurs in Purulia, but famine and death from starvation are mostly uncommon. This, I was told, is because of the active role taken at the local level.

Kalahandi is one of the tribal-populated districts in Orissa and, when I was working in Rourkela, Orissa, in the late 1970s and early 1980s, I passed through the Kalahandi district – but drought, famine and starvation were not a common sight. So, during 1996– 2000, I was surprised to hear the news that there were 'starvation deaths' in Kalahandi. During that period, there were three successive droughts in that district, resulting in shortages of food.

This had some impact on the health of local people in the form of malnutrition, under-nutrition and starvation; although I was told that the occurrence of starvation death was unlikely and debatable.

After the collapse of the former Soviet Union, in August 2001 Tajikistan was reported as having shortages of grain caused by two years of drought. The International Federation of Red Cross and Red Crescent Societies warned that a million people could face starvation unless they received emergency food supplies.

Not long ago, in August 2002, I read that: 'East China's Shandong Province is experiencing its worst drought in one hundred years', and in 1997, North Korea blamed severe drought and massive floods for its country's food shortages.

In the Middle East and some parts of Africa, the heart of ancient civilisations, drought still occurs as it used to in ancient times. However, they cannot avoid the effect of famine because of the countries' political systems, religious or tribal conflicts and war embargoes or sanctions, although some of them are rich with oil money.

It seems that drought can occur anywhere in the world, whether the country is capitalist or communist, dictatorial or democratic, rich or poor, developed or underdeveloped. The question is, 'Is it responsible for famine and starvation in the modern day world?'

The world population is increasing, but arable land is decreasing. Can the world produce enough food to feed its expanding population?

According to the WHO, 'Some 40,000 hunger-related deaths occur every day, mostly in rural regions' and 'millions of children do not get enough food to fully develop mentally and physically.' According to UNICEF, 'No less than half of all children under the age of five in South Asia and one third of those in sub- Saharan Africa, as well as millions in industrialised countries, are malnourished.'

The question remains to be answered: 'Is it possible to prevent food crises, hunger and starvation on our planet?'

Here I would like to quote a paragraph from one of my father's books published in 1970:

In an age when man has walked on the moon and is reaching still farther into space, he has yet to conquer the destitution, hunger, ignorance and diseases which afflict two-thirds of his fellow men on his own planet.

From Marx to Mao by D N Dasgupta

Hunger-related ill health and death are most probably due to poverty rather than famine. Culture, beliefs and lack of dietary knowledge are also contributing factors.

According to the Nobel Laureate Amartya Sen, the free press, the democratic political structure of the country and public action can prevent famine; he gave the example of India. Sen claimed that India had prevented famine since independence in 1947 because of its democratic institutions.

So, I have to admit that starvation death arising from famine, war and other natural disasters might be rare today; surely it would gain worldwide coverage by the global media.

However, Sen's logic was criticised by some, as it has doubtful practical applications with respect to starvation, when applied to some other countries of the world.

It is certainly true that the media, democracy and good infrastructure (that is, improved communications and distribution systems) have helped, and more scientific advancements in food production, storage systems, better healthcare and public action or famine relief have all played a role in the prevention of starvation. Unfortunately, starvation can result from war embargoes, sanctions, government corruption, trade policies, environmental degradation, overpopulation, International Monetary Fund (IMF) issues, economic oppression, mismanagement of food supplies and failure to provide better health care and famine relief. Until these are looked into, and the rich are willing to share the world's wealth, malnutrition, under-nutrition and starvation, death cannot be avoided.

A group of seven countries comprising the world's largest industrial market economies came together in 1976 to form the G7. The finance ministers of Canada, France, Germany, the United Kingdom, Italy, Japan and the United States of America meet several times a year to discuss economic policy.

The G8 consists of Canada, France, Germany, Italy, Japan, Russia, the United Kingdom and the United States. They meet every year for what is mainly an economical and political summit, and its presidency rotates amongst member countries each year.

The meeting of the G7 finance ministers in June 2005 was held at Gleneagles in Scotland, and focussed on debt, aid and trade. It was a major breakthrough in Britain's campaign for Africa. Many African countries' debts were written off and also development aid was increased. The eighteen countries that qualified for immediate debt cancellation were Benin, Bolivia, Burkina Faso, Ethiopia, Ghana, Guyana, Honduras, Madagascar, Mali, Mauritania, Mozambique, Nicaragua, Niger, Rwanda, Senegal, Tanzania, Uganda and Zambia.

When my wife and I went to Bolivia – one of the two countries in South America on this list – we took some clothes to donate to poor people and beggars. However, we did not come across many beggars in comparison with other poor countries in the world, where they tend to surround tourists. In Peru, we saw people trying to sell items, however small they were, to tourists instead of begging. The only beggars we saw were a few elderly women in front of the cathedral in La Paz, the capital of Bolivia. As my wife could not find anyone to donate to, we decided to give the garments to our local guide in La Paz, so that he could distribute them to the people in his village. He accepted these donations gratefully.

Money has to be better spent, with an urgent need to feed the starving, treat the sick and build better infrastructures. The poor nations need to aim for self-reliance. Cancelling debt will not solve the long-term problem.

It is good news that the poorer countries are going to agree to stamp out corruption, and world leaders are also agreeing a plan for reform of the World Bank and the IMF.

Drought can push prices and inflation up, sending millions of people into poverty and hunger. So too can climate change-related disasters.

Since 2008, the financial and economic crisis has spread across the world. To tackle these, the G20 was formed in view of further strengthening the international co-operation. The G20 is the group of twenty finance ministers and central bank governors from twenty countries, including Argentina, Australia, Brazil, Canada, China, European Union, France, Germany, Italy, India, Indonesia, Japan, Mexico, Russia, South Africa, South Korea, Saudi Arabia, Turkey, UK and USA. In addition there are five institutional representatives or members who can attend or participate in the meeting. They are the Managing Director and the Chairman of the International Monetary Fund, the President of the World Bank, the Chairman of the Development Committee and a member of the International Monetary and Financial committee. The G20 was established in the year 1999. The member countries cover two-thirds of the world population and collectively it comprises eighty percent of world trade and eighty-five percent of global gross national product. It is the forum for co-operation and consultation to promote financial stability in the international financial system.

The G20 group has now replaced the G8 group as the main economic council of the global wealthy nations. In April 2009, at the end of the G20 London summit, a package of US$100 billion and that the IMF will raise US$6 billion by selling gold resource to increase the the prospect of lending.

It seems that, although drought is still prevalent throughout the world, rich, developed countries like the USA do not report famine. Certain countries have also avoided famine by importing food and distributing it quickly and efficiently.

However, some countries in Africa, Latin America and Asia have still failed to overcome the problem of famine in the modern world in which we live.

Cattle Carcasses, famine in Ethiopia, 2006

Chapter Twelve
Accident and Accidental Disaster

Chernobyl, Ukraine
(Photo by Jason Minshull)

According to dictionaries, 'accidents' are unexpected and unpremeditated events causing damage to persons or to structures. An 'accidental disaster' is defined as the occurrence of a sudden incident or accident which disrupts the basic or normal function of society or community. This causes serious losses and usually affects the community such that it is unable to cope with its own resources. A prime example of this is the Bhopal gas disaster in India, which was one of the worst industrial catastrophes in the world.

Twenty-five years ago, just after midnight, toxic gas from the Union Carbide pesticide plant swept over the city of Bhopal and killed 3,800 people instantly. It was a blanket of toxic gas that formed an enormous cloud, which spread rapidly and widely, engulfing the city. Twice as heavy as air, methyl isocyanate (MIC) made up the base of the gaseous ball and, above this, there were successive layers of other gases: phosgene, hydrocyanide acid and monomethylamine with ammonia. This caused some people to go almost blind, and some shouted that their eyes were burning *chilli.* People were coughing, spitting and suffocating, and many choked and spewed blackish clots. Some shouted, 'I can't breathe.' Muslim women who wore veils or *burkhas* came out with their faces uncovered. In some places, weddings and religious festivals were taking place. Many people collapsed. People shouted, '*Bhago, bhago*' ('Get out of here').Some yelled, *'Bachoo, Bachoo'* (Rescue us, rescue us). Gases that smelled of boiled cabbage burned the eyes and suffocated the atmosphere, forcing people to flee in panic, mostly in the direction of the station – but they found that no train was leaving or coming to the station. There was chaos in the city and everybody wanted to run away, by whatever means were available. The most affected areas were the shanty towns, or slum areas, near the station. In desperation, some sought refuge in temples, mosques, churches and local houses. Hundreds of gas victims, alive or dead, poured into the local hospitals.

The local hospital was over-stretched, and a field hospital was set up in front of the local medical college and hospital. Whether victims were dead or alive, the local doctors and medical students did a tremendous job with limited means. Post-mortems conducted by two pathologists at that time suggested that hydrocyanide acid, one of the breakdown products of MIC, was responsible for killing the gas victims. Hydrogen cyanide was also found in Bhopal air and water on analysis.

The human tragedy of the Bhopal disaster has been vividly described by Dominique Lapierre and Javier Moro, in their book entitled '*It was five past midnight in Bhopal*'. It seems that the most dangerous toxic gas was the MIC, and more than twenty tonnes of this gas leaked into the atmosphere on the night of 2 and 3 December 1984, injuring

570,000 people, killing men, women, children and animals, such as cows, buffalo, goats and dogs, although small birds and hens escaped poisoning. However, since then, 16,000 more people have died and the final estimation was 20,000. It seems that the nature of toxicity and properties of the chemical are such that it affected the humans and certain animals, but not small birds. This might affect worldwide issues on animal experiments. MIC might be neurotoxin, and the long-term effect of the gas on the Bhopal community is still unclear. There are ongoing disputes with Union Carbide over the compensation of £250 million, which was based on £1,250 (100,000 rupees) per death, £625 (50,000 rupees) for permanent disability and £300 (25,000 rupees) for injury.

This incident made me frustrated and angry because, looking back down memory lane, I found that if I had been offered a job and had been working at this plant before the incident happened, I could have played a part or some role in the prevention of this disaster. A few months before the accident in 1984, there was an advertisement in the national newspapers in India: Union Carbide wanted an occupational physician to work in their pesticide plant at Bhopal. I was, at that time, in India, working at SAIL, based in Rourkela. I was interested in this job because of my interest in neuropathy and pesticides. I was on the verge of completing some research on the neuropathy of miners working with vibrating tools. I thought it was high time to move on to the chemical industry, which deals with pesticides. This was important as, in my dissertation work for my MSc at London University in 1976, I pointed out the high risk of the development of polyneuropathy among the agricultural, forestry and fishing industries in northern England. The possible explanation of this association might be industrial exposure, although in my discussion I did mention that, as a result of the high standards in industry today, toxic neuropathies due to industrial substances are very rare. However, if any occur, it is mostly due to accident or substances new to the industry. Probably this statement is rightly applied to the Bhopal tragedy.

I applied for the job at Union Carbide and was called for an interview, well before the disaster. I did not get the job and the administrators did not encourage me to go for the post. Prior to my interview, I requested that they show round me the plants, especially the hazardous areas and the medical centre, but the administrators were reluctant to take me round. However, they did vaguely mention the chemical storage tanks and the medical centre; they did not show me the medical centre, but I was introduced to the lady doctor who was working there at that time. We did not have much conversation and the only discussion I had – as far as I remember – was that they asked me about medical information on certain chemicals, including MIC, and their emergency treatment. I wondered whether the company had a proper occupational health department or not. These people discouraged me from joining the company. I felt strange and now realise that there must have been some problems, most probably health and safety issues, which were building up there and resulted in such a catastrophic disaster.

After the tragic accident in Bhopal, some progress has been made on our understanding of the toxic effects of MIC. Most findings are on the respiratory and reproductive systems. Birth defects and retarded growths have also been reported There is very little on neurological disorders, except some neuromuscular dysfunction. Some of the studies were done on the basis of animal experiments. Vivisection is a big issue nowadays in some countries of the world because of the animal cruelty involved in some animal experiments. There is also a debate on the validity of testing on animals and human reproducibility, or vice versa.

In April 1986, there was an accident at the Chernobyl nuclear power plant in the Ukraine, which destroyed one of the reactors by fire and explosion. This accidental disaster immediately killed thirty people and 209 people were treated for acute radiation poisoning. Large areas of Belarus, the Ukraine and Russia were contaminated, and more than one million people were possibly affected by radiation. Initially, 45,000 residents were evacuated from within a ten-kilometre radius of the plant. Later on, a further 116,000 people, living within a thirty-kilometre radius, were evacuated and subsequently relocated. Even years after the accident, a further 210,000 were resettled into less contaminated areas. In late 1995, the WHO linked 700 cases of thyroid cancer among children and adolescents to living in the contaminated areas. There was also an increased incidence of birth defects that left children with malformed or absent limbs due to congenital limb defects.

For a considerable period of time, there have been serious issues of congenital limb defects and their association with environmental or occupational factors throughout the world. I was interested in congenital limb defects and those associations, so, when I was working at the Portsmouth Disablement Centre in the UK, I looked at the possible link between congenital limb defects and various environmental or occupational risk factors on those people who attended the Centre. Naturally I looked at the clusters of congenital limb defects surrounding the British nuclear establishments at Aldermaston and Burghfield in Berkshire, but I could not find any clusters near those establishments, although the study was limited and the numbers were small. However, there is no doubt that a nuclear accident like Chernobyl gives great insight and responsibility to the nuclear industry, and shows how important it is to protect their employees, the community and the children.

The world's worst offshore oil disaster took place in the evening of 6 July 1988 on the Piper Alpha drilling platform, 193 kilometres off the north-east of Scotland, killing 167 men. The Piper Alpha platform is the largest and the oldest platform in the North Sea oilfield. Oilrig platforms in the North Sea are mostly built onshore and the structures are then carried out to the North Sea, where the platform and the oilrig are assembled and constructed. In the1970s, when I was in north-east England, I met some men who were constructing the platforms and I asked them whether the platform would withstand bad weather or hazardous conditions. Generally, it has – although, since drilling began in the North Sea in the 1970s, there have been 300 deaths on Britain's oil installations, often accidents caused by bad weather. However, nobody expected an offshore oil disaster such as the gas leak on the Occidental oil drilling platform, Piper Alpha, which caused an explosions and fires, killing 73.89 per cent of the total of 226 men on board. I gather that, since then, there have been a lot of changes to health and safety procedures in order to prevent such disasters on oilrigs.

In the USA, an off shore oil rig explosion took place on 20 April 2010 in the Gulf of Mexico, killing 11 people and injuring many of 120 crew who jumped overboard; 42,000 gallons of oil a day leaked into the sea covering 400sqare miles. The oil rig platform burned for a day before it sank. The oil slick reached the coast lines and 1000 km cost lines have been affected, in spite of a great effort to stop leakage. This had catastrophic effect on marine and wildlife, as well as the local fishing and tourism industries. This became the worse environmental disaster in USA's history. However, after five months the British Petroleum was finally able to seal the well, which caused four million gallons of oil spill in the Gulf of Mexico.

There are 3,350 platforms along the coast of the Gulf of Mexico. Accidents on the oil rigs are not uncommon; since 2001, 69 people have died, 1,349 injured and 858 fires and explosions occurred in the Gulf of Mexico.

However the largest explosion occurred in Europe in recent times on 11 December 2005. It was not an offshore incident but an onshore oil storage terminal disaster at a Hertfordshire depot in Britain.

Within half an hour there were three large explosions possibly due to fuel-air explosion. The fire lasted for four days as twenty silos burned. Twenty-five fire engines, twenty support vehicles and 180 fire fighters were involved in controlling the blaze. Known as Buncefield fire, the incident took place near the M1 motorway in Hamel Hampstead, Hertfordshire, England. The blast and fumes produced black smoke and affected the surrounding residential area and offices. Hundreds of homes were evacuated and the schools and libraries within a ten mile radius were closed. The cloud of smoke also reached south-east England, the English Channel, northern France and most probably northern Spain. The small particles in the smoke were hydrocarbon. The carbon monoxide, nitrogen oxide and ozone concentration were found to be low as soot particles were the major components. There was a question of groundwater pollution though later on drinking water pollution was excluded.

There were no fatalities but forty-three persons had injuries with two considered to be serious. However, a total of 244 people required medical aid.117 people including members of public complained of incident-related symptoms. Sixty-three members of the emergency services had respiratory problems and fifty percent of them had sore throats. According to the Health Protection Agency, UK, there has been little in the way of short-term adverse health effects but the long term effects still need to be properly monitored, although up to now no major after-effects have been reported. However, the Health Protection Agency and Major Investigation Board in the UK have highlighted the safety measures necessary in the future with respect to preventing the escape of such fuel, which forms a flammable vapour that pollutes the atmosphere.

Another major fuel related fire and explosion occur recently in India. This occurred at the Jaipur Oil Depot in Rajasthan on 29 October 2009, killing twelve people and injuring more than three hundred. The fire lasted for a week. This accidental oil blast took place when petrol was being transferred from a giant tank to a pipeline at Indian Oil Corporation's oil depot.

In 1961, I was a first-year medical student in India. In that year, at the age of eighteen, I had an accident in the chemistry laboratory while experimenting with acid. Before the accident, my vision was 100 per cent perfect and I never wore glasses. During the experiment, students did not wear safety glasses; at that time, there was no provision or education for the use of safety glasses. My face, including both eyes, was burnt and the ophthalmologists were not sure whether they would be able to save my eyes or not. However, after forty-eight hours, it was found that my left eye was saved but the right was damaged. Many people were shocked by this dramatic incident. Since then, I have had various treatments in India and the UK, including cornea grafts for my right eye. I had some success with my right eye's vision initially, but it has gradually deteriorated and now I am completely blind in my right eye.

This incident affected my whole life, and my lifestyle was also changed. It affected my education, sports, hobbies, career, employment, family life and health. I was more introverted and probably became less social. I dealt with people who asked me, 'Were you born with your right eye like this?' Some showed sympathy when they learned that it was the result of a laboratory accident, and some commented that the evil star had taken my right eye away. Some astrologers predicted that my left eye was also in danger and, in order to prevent such consequences, I should wear gemstones. Some said that my accidental suffering was the result of my sins. I told them that I would have nothing to do with such awful and unnecessary remarks. Throughout history, it has been shown that scientists, inventors, explorers and adventurers suffered from accidental injuries, diseases and death for the betterment of mankind; the accident did not mean it was the result of any sin. It was more important to learn how to prevent this. Despite some setbacks and difficulties, I was able to lead a normal life and carry out my professional qualifications, job and duties; I married and became the father of two daughters, and became a grandfather.

Eye injury is one of the most common injuries at work or in industry. However, sports and recreational eye injuries and injuries associated with fireworks are also not unusual. One of the most common eye injuries among children and adults arises from sports activities. 90 per cent of eye injuries are preventable. The provision and use of safety glasses and goggles are important. To prevent laboratory accidents like mine, the use of safety glasses should be compulsory for students in schools and colleges throughout the world. This is the case in countries like Britain, but I would not be surprised if it is still lacking in India and other countries, where safety is not such a priority.

Up until now, I have never worn gemstones on the advice of an astrologer or others. Instead, I did some investigation on gemstones' mystical powers and healing abilities. For that, I went to the Himalayas, Sri Lanka, South America, China, Kashmir and other parts of India, and various volcanic areas, looking at various stones and gemstones. One of the greatest collections of gemstones and volcanic rocks I have ever seen is in the National Museum in Prague, the capital of the Czech Republic. Since ancient times, there have been numerous myths and legends associated with gems. Some tell of crushed stones, others of stones with special powers of healing or that protect or give good luck to the wearer. Gems were used in the form of powder or mixed with other medicines in ancient India by Ayurvedic doctors called *vadha*, and later on during the Muslim period by doctors called *hakeem*, and those practices are still in vogue in the Indian subcontinent. In ancient Greece, physicians realised the medicinal value of gemstones and used gems as a part of their treatment.

Volcanic eruption, magma and crystallisation are important factors in the formation of some varieties of gemstone. Peridot (olivine) is a type of gemstone which is found in Hawaii and is formed by crystallisation of rock at high temperatures. The black lava gravel is dotted with green crystal fragments of olivine and we noticed this when we visited Big Island in Hawaii in March 2008. Three miles east of South Point there is a green sand beach which is famous for olivine crystals and is formed by the collapse of cinder cone. South Point is not only the most southerly point of the Hawaiian island but is also the most southerly point of the USA which we visited.

While there, I also saw jewellery made of peridot crystal taken from the green sand beach being sold at the local market in Kona and I brought a piece as a souvenir. In 2003, peridot was also discovered on Mars, making it the first gemstone on Earth to be found elsewhere in our universe. Peridot is believed to have some medicinal value and has been used to cure asthma and reduce fever, as well as for highly practical purposes such as for goblets in which to keep medicine. I looked at the chemical composition of olivine and found that it usually contains mild silica, iron and magnesium.

Crystal healing was a part of the rituals of medicine men in ancient tribes. Even today, crystal healers in the UK and other parts of the world believe that each stone or gem has the power to influence the health and well-being of specific parts of the body. The stones are placed on vital nerve points and the light reflected from stones is absorbed by the body, with healing energy. The composition, crystal structure and optical properties of the gems, including the refractive indices (RI) and birefringence, might play some part in healing activities, but I am not sure about the scientific explanation of protection from evil. I looked at some such protective stones:

1. Agate (chalcedony) is believed to guard against danger or evil, and is believed to be excellent for the eyes.
2. Amber is said to be a healing stone for the eyes and brings good luck.
3. Aquamarine protects against the will of the Devil.
4. Diamond is said to protect from evil.

5. Emerald drives away evil spirits and protects from accidents.
6. Garnet protects from harm.
7. Jasper drives away evil spirits.
8. Mandarin is said to protect the wearer.
9. Opal protects from the evil eye.
10. Peridot is said to protect from evil spirits.
11. Tourmaline is said to protect the wearer against any danger and misfortune.
12. Turquoise is said to represent protection against violent accident and danger.
13. Gomedha ensures natural safety and protection from the deadliest enemy.
14. Cat's eye conquers the enemy and offers protection from an enemy.
15. Ruby protects against disaster.
16. Pearl provides long life, and all sins are said to be washed away.
17. Coral protects from evil spirits.
18. Topaz (yellow sapphire) protects against evil spirits.

It is very difficult to find any scientific meaning or logic here, and so I looked at Indian astrology and found that in astrology there was an important role for gems.

This tradition says that the use of gems affords protection from the evil influences of planets and enhances their beneficial effects. Gems contain divine power. The rays come from the planet and are filtered by gems to obtain vibrations for the human body. There are good and bad effects of gems. A general belief about birth stars, planets and gems is that they are considered to be lucky or important for people born under their influence.

There are 12 birth stars which are Aries, Taurus, Gemini, Cancer, Leo, Virgo, Libra, Scorpio, Sagittarius, Capricorn, Aquarius, Pisces and the ruling planets are ten that is Mars, Venus, Mercury, Moon ,Sun, Jupiter, Saturn, Pluto, Uranus and Neptune. The gems are Ruby, Coral, Diamond, Malachite, Emerald, Pearl, Agate, Rose-Quartz, Fire-opal, Amethyst, Garnet, Sapphire, Chrosobery and Green-beryl.

The Chinese also believe in astrology and that has a long history. In Chinese astrology, there are twelve zodiac signs, which are: Rat, Ox/cow, Tiger, Rabbit, Dragon, Snake, Horse, Goat, Monkey, Rooster, Dog and Pig. An individual sign is based on the year that person is born.

However all these have not influence my opinion of whether or not to wear gem stones

Trains have always attracted me and I remember, during my childhood in India that I used to go to the local small railway station, stand on the railway bridge and watch how the steam engines were passing, making hissing sounds, full of smoke, dust, steam and fumes. I spent some of my vacations at my maternal aunt's place at Patna and their house was near the Calcutta–Delhi mainline. Every time a train passed through, I used to run to the veranda and look at which train was passing and how fast it was running. One of my cousins was very good at spotting trains. Without seeing the train, by judging the time and speed, he could tell me the name of the trains going up and down the line. At that time, I was rather jealous of him and thought how clever he was. However, the irony is that, a long time after that, I escaped a fatal accident on that mainline near the Allahabad station while I was travelling from Calcutta to Delhi on one of the Indian super-fast trains. Of course, in India most of the engines are now either electric or diesel. Most of the steam engines have thus been replaced, and the production of steam locomotives in India stopped in 1972.

The invention of steam engines and the railway system was a great achievement of science and technology during the age of the industrial revolution. The great invention of the modern steam engine was made by James Watt (1736–1819) at Glasgow in 1765, and the birth of the steam-powered locomotive revolutionised the transport industry. It was possible because of men like Richard Trevithick (1771–1833) and George Stevenson (1781–1848). Richard Trevithick, in south Wales, constructed the world's first steam locomotive in 1803. George

Stevenson formed a locomotive company in Newcastle, which was the first locomotive company in the world. The first modern locomotive train ran from Darlington to Stockton-on-Tees in 1825. The work started in 1822, and it was opened on 27 September 1825. There was a twelve-mile track between Darlington and Stockton, and it took George Stevenson and his colleague two hours to cover the first nine miles. I was very excited when I first saw the museum piece of this locomotive at Darlington station. When I used to live in that part of the world, I used to go to Yarm and watch the bridge over the river Tees where George Stevenson and his colleague had driven the first locomotive of the world successfully. This event was well recorded in watercolour paint.

Twenty-seven years later, the first train in India ran between Bori Bunder and Thana on 16 April 1853, and this was a twenty one- mile rail track. The Great Indian Peninsula Railway Company was responsible, and its first director was, again, George Stevenson. The history of the Indian railway is fascinating: how the engineers and workers built the railways, bridges and one of the greatest networks that connected the whole of the Indian subcontinent. During the construction stage, there were incidents where men lost their lives or were attacked by animals. The classic example of this type of incident was the construction of the Bengal Nagpur Railway in 1890, in the Sarandra forest near Chakradharpur in eastern India, where herds of elephants were run over and killed. I gather that the tusks of those elephants were kept in the director's office in London. In India, each time I passed through the Sarandra forest, the incident always came to my mind.

After Indian independence in 1947, the Indian railway became part of the public sector, and I gather that the Indian railway has over 63,000 kilometres of track at present but has got poor safety records.

The number of accidental deaths in India due to railway accidents is one of the highest in the world; every year, nearly 300 railway accidents occur in India, although the number of crashes declined in recent years (about 177 in 2008-2009)

However, I had never been caught in a major railway accident in my life, until 10 October 1977, when the super-fast train in India on which I was travelling crashed at Naini station, near Allahabad. This incident still from time to time disturbs my mind. It was 3 a.m. and I was dozing in my air-conditioned compartment. I suddenly woke up at the sound of a loud bang, followed by a big jerk and the immediate halt of the train. The lights and air conditioning went off. I looked through the window; it was pitch dark everywhere. My seat was near the main door of the carriage and so I immediately came out of the compartment, and was horrified at the sight of the front compartments. I was extremely lucky, because the carriage in which I was travelling was safe, but all the compartments in front of us either were on top of one another or had derailed and fallen sideways; everywhere, people were screaming and banging. The trapped passengers tried to break open the doors and windows. I walked to the front, towards the collision site, and saw that the engine was completely smashed and some people were trapped underneath, but it was so dark that I could not see fully. I could only hear moaning and groaning like, '*Mar giya, mar giya*', which means 'I am dying, I am dying'. This was the first time as a doctor that I felt helpless, and I suddenly realised how difficult it is to rescue victims who are trapped, rather than actually giving treatment. I was not sure how the accident happened, but I was told that a goods train was stationed at the Naini station and our super-fast train, which was travelling at high speed, hit this goods train. A signal failure might have been the reason for the tragic incident.

I was not sure what time the railway rescue team came, but I suddenly realised that it was dawn and I was worried how I would reach Delhi airport, from where I had to catch the midnight flight to London.

I had come for a week to India, and my younger brother Aloke - who came to Howrah station to see me off – did not expect this.

When I first got onto the train, I was surprised to see somebody occupying my reserved seat. However, Aloke helped me to relocate to a seat opposite my allocated spot, which later prevented me being thrown off the seat. Aloke was working as a sports editor for a Calcutta-based newspaper. My immediate family, my wife, daughter and my parents were at that time in Teeside; my wife was expecting our second child.

Some fellow passengers told me, 'Let's go to the station and find out how long we have to wait for the next train or any other alternative mode of transport that is available.'

Another person stated, 'I have to make a telephone call or telegram our relatives, otherwise they will be worried.'

We all decided to walk down to the local station, carrying our luggage – I was worried about losing the luggage from our compartment. When we stepped onto the platform, I saw some dead bodies that had been removed from the accident site. I saw that the station and platform were crowded, and it was very difficult to make any outside communication from there until I had waited for a considerable time.

The authorities informed us, 'Until this line is cleared, no train is able to pass through, and we do not yet know how long it will take to clear the line.'

I asked, 'I have to go to Delhi to catch a flight – how can I get there?'

One of the railway police officers told me, 'The authorities are providing a lorry for passengers that will take you to Allahabad station, which is only three miles away, and from there you can get a train for Delhi.'

I caught the lorry and reached Allahabad station, but the station authority at Allahabad informed me, 'No trains are running from Allahabad station because of this accident.' The authorities also told me, 'Trains are running from Kanpur, and to catch the train from there you have to go to the local bus stand and take the bus for Kanpur, and from there you will get a train for Delhi.'

I took the first bus, which was just after 6 a.m., and reached Kanpur by 9 a.m. My youngest sister, Leena, and her family lived in Kanpur, so I went to her place and found that my sister and her husband were not at home; they were at that time in Lucknow. However, one of their business employees took me to Kanpur station and put me on the Assam–Delhi train. I also asked him to inform my brothers in Calcutta that I was safe following the incident. Without further event, I reached Delhi station at 8 p.m., from where I took a taxi and reached Delhi airport well in time for the midnight flight to London. From the airport, I tried to contact by telephone my brothers in Calcutta and my wife in Teeside. I was able to contact my wife in England, but not my brothers in Calcutta. However, over the phone I did not tell her about the accident.

I was waiting at the airport lounge to board my London flight when I suddenly heard an announcement that I was wanted at the desk. I went to the desk and found that I had a telephone call; the caller was one of my distant relatives. She informed me that everybody in Calcutta was worried, as I was on the missing list following the accident. She was happy to hear my voice and informed my relatives in Calcutta about my safe departure. Although I reached England safely, I felt guilty that circumstances had not allowed me to do much for the accident victims at the time. I also realised that, in any disaster planning, it is important to rescue the victims in a difficult situation rather than give medical aid, which might be a secondary issue in certain circumstances. In India, the rescue process is slow and outdated. Most of the time, it is usually dependent on individual or public effort rather than an official one, which is usually delayed. Sometimes, apathy or indifference from some of the public regarding rescuing victims is not uncommon. The method of rescuing victims is different in India than the method we see in the Western world.

Of course, this was nearly thirty-three years ago and the situation must now have improved from the former method of rescuing victims.

In the past ten years, there has been an increase in the number of rail disasters in UK. In between 2000 and 2009 there were 31 railway accidents and some people are blaming privatisation for this.

The prime examples are Potters Bar in 2002 and the Hatfield rail crash in 2000 and, in 2004, the Berkshire rail crash, where some of the victims were treated in the North Hampshire Hospital in Basingstoke, the hospital in which I used to work. Our hospital staff did a very good job, although our department was not involved.

In other parts of Europe, there has also been an increase in railway accidents and, as in Britain; the majority of the accidents took place on unmanned level crossing areas. Considering the various rail disasters of the world, I cannot say whether they are part of the public sector or private sector, but rail disasters are still dramatic and have sad consequences, the majority of which are preventable.

The world's greatest and most remarkable shipping disaster was the Titanic in 1912. This was in the pre-flight era, when the big ocean passenger liners crossing the Atlantic were a common phenomenon. On 10 April 1912, the Titanic set sail from Southampton on her maiden voyage to New York. At that time, she was the most technically advanced, largest and most luxurious ship ever built. At 11.40 p.m. on 14 April 1912, she struck an iceberg about 400 miles off Newfoundland and 1,000 miles east of Boston. Less than three hours later, the ship disappeared below the sea. Only 705 of the 2,227 people on board survived.

Some of the survivors told the tragic story of their sorrow at the incident, which were recorded in books and more recently in the Oscar-winning film. I first became aware of this in my childhood from the mouth of one of my friends who had read the book; later on I also did. This topic came out of our discussion on icebergs. Icebergs are big chunks of ice, which break off at the end of glaciers and drift out to sea. Icebergs are very dangerous to ships and boats.

The Titanic hit such an iceberg and sank in the North Atlantic.

When I was in New York City in 1997 and took a boat trip on the bay surrounding New York City, the guide was showing us the Statue of Liberty, Ellis Island and Pier 54. Pier 54 was where the victims' bodies from the Titanic were brought ashore, and the survivors passed through it without the immigration formalities at Ellis Island. The remains of the Titanic are now located on the ocean bed, and were found in 1985 by Dr Robert Ballard, oceanographer and marine biologist.

Since the discovery of the new world, many ships and aircraft have been lost in the Bermuda Triangle. The Bermuda Triangle is a region of 440,000 square miles of open water, south of Bermuda.

This is the area where some of the Sargasso Sea, part of the Atlantic Ocean, flows into a triangle, and some people believe that supernatural things happen there. Rogue waves, methane bubbles from the ocean bed, and so on, have been blamed, but a satisfactory explanation is still lacking for these unusual disappearances. There are debates on the science behind it, resulting from a manufactured mystery due to misconceptions, or inadequate research and sensational storytelling. Once, I was discussing this topic with an author of the book entitled, *Mysteries on the High Seas* who had done some research on lost ships in the Bermuda Triangle and, according to him, the Bermuda mystery and ship disappearances have been exaggerated. He thought that there was no real evidence that the seas around Bermuda were more dangerous than other parts of the ocean. The author was Dr Phillip MacDougall, and he wrote about this in his book However, my own feeling is that this part of the world lies on a hot spot of the Earth, near or along the Mid-Atlantic Ridge or plate boundary. Hurricanes, tornadoes, ocean surface currents, volcanic eruptions, emissions of toxic gases, earthquakes and tsunamis are common occurrences. Moreover, Earth's strange electromagnetic phenomena might play some role. If ships or aircraft are caught in such environments, no doubt there will be losses or major disasters.

Rail, road and marine transport are most often used for the transportation of hazardous material in the form of liquids, gases and solids, and there will on occasion be explosions, collisions and spillage, which might give rise to many types of transportation accident and disaster. The most common road disasters are when oil or gasoline tankers explode. Such an incident took place in Spain in July 1978, killing more than 120 people, and in Afghanistan in November 1982, which killed more than 2,000 people. In November 2000, more than 150 people were killed in Nigeria when an oil or gasoline tanker hit cars and exploded.

My first experience of this type of incident was when I was a final-year student in Calcutta. I saw victims with tar, pitch and bitumen burns when an overcrowded double-decker bus hit a roadside boiling-hot bitumen carriage. Road repair and tar surfacing was going on, and the victims were the passengers on the bus. This type of incident is usually preventable if appropriate precautions are undertaken. In the Western world, during this type of repair or construction work, the lanes are usually fully closed; the traffic is diverted, and it is done at night when the traffic is minimal. Of course, nowadays traffic jams are more problematic, and I have also noticed over the years in the UK that road construction or repair work occur more in daytime and less at night. No wonder there are more traffic jams, bearing in mind that the volume of road traffic has increased enormously recently. Moreover, sometimes the diverted traffic signal is so badly signposted that motorists get lost on the way, resulting in unnecessary harassment by the police.

Marine transportation is responsible for fire, explosions and the burning or sinking of vessels on the sea, and the most environmentally hazardous issue is oil transportation by big tankers. Every year, oil tankers transfer 1,800 million tonnes of crude oil around the world by sea. Since the introduction of big tankers, there has been an increase in the number of accidents and oil spillages.

In December 1987, more than 4,300 people died when a ferry collided with an oil tanker near Manila. Oil tanker accidents off the UK coast have also increased. From time to time, there is big news about environmental disasters due to the oil spillage from big tankers. The worst incident off the UK coast was in 1967, when a tanker ship named Torrey Canyon spilled her entire cargo of 120,000 tonnes of crude oil into the English Channel. As a result, 270 square miles of sea was contaminated by an oil slick and ninety-three miles of Cornish coast was affected.

The disaster that took place off the Spanish coast on 13 November 2002 was the most costly maritime accident in history. The tanker Prestige, which held 70,000 tonnes of fuel, sank into the Atlantic Ocean, spilling 20,000 tonnes of oil. At least 100,000 to 200,000 birds suffered the effects of the Prestige's black tide, which extended from the Spanish coast, north to France and south to Portugal. It was estimated that 75,000 tonnes of waste were produced, even eight months following the incident, which included fuel, oil-covered sand, seaweed, dead animals, and so on. There was also a question of how to remove the rest of the fuel from the sunken ship, which many believe will continue to leak, even twenty years after the incident.

In 2004, there were two major rail transportation accidents, which are worth mentioning. One was on 22 April 2004 in Ryongchon, North Korea, when 1,300 people were injured and 154 confirmed dead after the station explosion. According to North Korean official sources, two railway wagons were carrying dynamite and exploded after getting snagged by overhead electric cables. The other accident was on 18 February 2004 in Nishapur, Iran, when runaway railcars loaded with petrol and sulphur products rolled down the rails, caught fire and exploded, killing more than 320 people and destroying five villages.

Most recently at midnight on 1 July 2009, a freight train carrying liquefied petroleum gas was derailed in the town of Viareggio in Tuscany, Italy, when the axle on the first wagon broke, resulting in the train running off the tracks and crashing into the houses as it exploded like a bomb. The flames spread to the entire street as buildings collapsed, killing some residents as they slept. Burns victims ran through the blazing street in the middle of the night. Thirty -two people were killed, twenty-six were injured and a thousand residents were evacuated.

The USA is a big country, well-connected by rail, road and air. Hazardous materials released from railroad tank cars are quite common. Since 1980, there have been at least ten major hazardous material-transport accidents in the USA, mostly involving cars and rail fleet.

There is no doubt that flying is one of the safest forms of transportation, but disasters still occur at frequent intervals. I looked at the major air disasters during the period 1990–1997 and studied the nature of them. They cover collisions in the air, midair explosions, crash landings, crash take-offs, collisions on the ground, bad weather and running out of fuel. Professionally or non-professionally, I have hardly come across any air disaster survivors, as survival from an air disaster is usually rare. When I hear of any air disaster, it always reminds me of the shocking incident on 14 December 1990, when the Indian airline A320 flight crashed while approaching Bangalore airport, resulting in eighty-nine fatalities. Among the victims were a just-married couple. We knew the bride's family very well. The bride's father was one of my engineer colleagues in a steel plant where I worked. They had two daughters, who grew up in front of us in Rourkela, India. The trauma that the parents suffered from losing the daughter and the son-in-law was very difficult to overcome.

The most common form of transport accident is the car accident. For the last decade, the number of cars per family has been increasing rapidly, and so too the number of cars on the roads. No wonder the car accident rate is high. In England and Wales between 1995 and 2000, there were 1,282,563 minor injuries, 191,870 serious injuries and 15,797 deaths as a result of road traffic incidents. It seems that, in recent years, there are not many people in the world who drive and escape some sort of vehicular accident. I am not an exception. However, I have never been involved in any major car accident, but I have seen a number of car-related traffic incidents and injuries in my professional life.

When I discuss car accidents, my story remains incomplete unless I talk about my near-miss accident on a German autobahn (motorway) in 1977. This was at the time of the Cold War and during the days of border control, before the establishment of the European Common Market and the reunification of Germany. Germany was partitioned into East and West by the Berlin Wall (it divided Berlin into Russian-controlled and American/British-controlled areas). For that reason, travel to Berlin at that time by road or car from West Germany was not comfortable, as one had to pass through several checkpoints, and there was a marked contrast in road conditions between East and West Germany. There were more pot holes, uneven surfaces and unrepaired roads on the East German side of the autobahn than in West Germany. The facilities along the autobahn were few, and I also noticed beautiful pine woodlands along the autobahn. These had been there since the time of Adolf Hitler and helped Germany to camouflage military movements at the time of the Second World War. During the fall of Berlin, fierce fighting took place between German and Russian soldiers in these woodlands. Some say that if one digs in the woodland now, one will still find corpses.

If any car broke down on that road, there was no breakdown service available to rescue the car and its passengers. This we realised when we were coming back from Berlin after a short holiday break. We left Berlin in the late afternoon and, after driving nearly 150 kilometres through East Germany, my car suddenly broke down on the autobahn due to an electrical fault that resulted in us being unable to drive further. We pushed the car to the nearest lay-by and were stranded there, as we were not able to contact the Automobile Association (AA) rescue service; no such service was available in East Germany.

Salil, my wife's cousin, tried to stop some of the vehicles in an attempt to get help, but he failed. We could not see any nearby place where we could ask for assistance. We all started to worry, as it was getting darker. At last, at around 9.30 p.m., Salil was able to stop a vehicle and request help from the driver, who kindly agreed to give us a lift by towing our vehicle up to the border of West Germany, where hopefully we could contact the AA. I did not have proper towing equipment but, fortunately, that gentleman had some gear that he connected to the front of my vehicle. The rope was made of steel wire with a fibre cover, quite strong enough to pull my vehicle along with all the passengers. All together we were five passengers: Salil, Marianne, my wife, my daughter and myself. My wife was expecting our second child. I was in the driver's seat at the steering wheel, and I never had such an experience in my life. The vehicle which was towing my car went at a speed of sixty or seventy miles per hour. To keep up with that speed on the motorway was not easy and required tremendous concentration, visual perception and co-ordination.

Sometimes the speed was so high that unconsciously I put my brake pedal down, causing some problems for our towing gentleman. As far as I remember, we travelled approximately one hundred kilometres, but the last bit was not pleasant as the steel rope broke abruptly. I suddenly heard the screams of Marianne from the back seat; there was a big articulated lorry behind us. To escape a fatal accident, I moved towards the right and stopped. When we got down from our car, we were horrified to see that the articulated lorry was standing six inches behind my car. The lorry driver had had to put on his double emergency brake to stop his vehicle, resulting in only a six-inch gap. Otherwise we would not have escaped injury and fatality.

After surviving such a potentially fatal near-miss, the next question was how to tow us the rest of the journey, as the steel rope was broken. There was no other spare rope to use. The only solution was to use the rest of the rope, which was now quite a short piece. Alternatively, a silk sari that was with us could be used as a rope. My scientific knowledge said that silk rope would be stronger than the steel rope and, if we made a silk rope by twisting the sari, we could easily use it. However, this was not required, as we were able to use the remaining steel rope for the rest of our towed journey. Soon after, we arrived at the border post, and we felt great relief as we reached West Germany. We were able to call the AA straightaway and await their arrival. We also called Marianne's brother, who was in Duisburg, and he drove all the way to our assistance. However, in the end, the AA was able to fix our car and we drove on that afternoon to continue our journey and reached our destination safely in Duisburg, where Marianne and Salil lived. The man who rescued us was a good man; from time to time, one finds a man whose deeds are difficult to forget.

In 2002, 1.2 million people were killed and fifty million injured in road traffic accidents (RTAs) worldwide. The cost globally was $518 billion. Regarding statistics of people killed in RTAs, Latin American countries are top of the list, followed by Europe and Asia. To prevent road traffic injury, more importance has to be given to safer roads, safer vehicles, better drivers and the reduction of motor vehicle traffic by safer alternative transport and proper land-use policies.

The police data on British road casualties from 1986 to 2006 showed that twenty mph zones were associated with a forty per cent reduction in casualties and collusions in London. This supports that the speed management is one of the key factors in the prevention of road traffic accidents and injuries.

Lastly, terrorists most often use cars, aircraft and railways as methods of killing innocent people as part of terrorist activities.

There is no doubt that the transport industry is responsible for being the major cause of injury or fatal accident. More research is necessary to improve and ensure people's safety in all circumstances.

Icebergs that cause navigation hazards

Chapter Thirteen

Human Movement

Emergency shelter for disaster refugees

The history of migration is almost the history of mankind. There is no record of when the first men appeared on this earth, but there is a fossil record which supports the theory that manlike creatures existed around fourteen million years ago.

Once, I was in a game reserve park where I was observing the behaviour of a chimpanzee. I was impressed by the way the ape was behaving and how it was similar to certain human behaviours. At the end, I saluted the chimpanzee. Some people standing next to me laughed at me. I saluted the chimpanzee as a mark of my respect to a common ancestor, as the ape's behaviour reminded me of the theory of human evolution and the common ancestry of man and the apes.

Man's far distant ancestor (*Ramapithecus)* was an apelike creature that could walk upright around 4.4 million BC and evolved in Africa. Scientists were able to reconstruct the sequence of human evolution by studying the bones of hominids. They believe that the human race underwent evolution first and then spread throughout the planet by migration. This is generally accepted by most scientists today, though the theories of human evolution and migration are still subject to further discoveries.

The first hominids (*Australopithecus*) evolved from apes in east Africa in about 4 million BC and were not as erect in posture as modern humans. The first human hominids, which are called *Homo habilis* ('skilful man'), evolved in Africa around 2.5 million BC. The first human species to leave Africa was *Homoerectus* ('upright man'), around 1.8 million BC; they migrated to Asia and later to Europe. *Archaic Homo sapiens,* whose human characteristics were shared by *Homo erectus* and *Homo sapiens,* appeared in 300,000 BC, and they may be the ancestor of the Neanderthals. In 100,000 BC, *Homo sapiens* (or modern humans), who were living in Africa, started to migrate to other parts of the world; in 30,000 BC, *Neanderthals* disappeared and *Homo sapiens* continued to spread across the earth. The *Neanderthals* were common in Europe and some parts of Western Asia. They lived in caves about 110,000 to 45,000 years ago. Until now 400 *Neanderthals* have been found. In Europe, skeletons were discovered in Belgium 1829, in Gibraltar in 1948 and in Germany 1856.

Neanderthals roamed over Europe until 30,000 years ago, when they suddenly disappeared. They were replaced throughout Europe by or possibly interbred with *Homo sapiens* (modern humans). *Homo sapiens* came into Europe from the east, replacing the *Neanderthal.* DNA studies suggest that *Neanderthals* were not our ancestor; however, they might be distant biological cousins.

Neandertal, a valley near Mettaman, is the site where the first species of homo neandallus was found in Germany. The Neandertal is a small valley of the river Dussel in the Germany Federal state of North Rhine-Westphalia, located about 17 kilometres (7.5 miles) east of Dusseldorf.

The Neandertal was originally a limestone canyon famous for its scenery, waterfalls and caves; however industrial mining during the nineteenth and twentieth centuries saw the extraction of most of the limestone. It was during the mining operation that the bones of the original Neanderthal man were found in a cave in 1856 and further excavations took place between 1997 and 2000.

When I visited the site in April 2010 there was no cave or cliffs which were long gone but the excavation areas are well marked with stones and poles and are still worth visiting. The natural surrounding the area is beautiful and the remnants of skeletons and stone tools are kept in the nearby Neanderthal museum which I also visited.

Homo sapiens came to America from Asia about 25,000 years ago across the Bering Strait. I saw the classical illustration of human evolution when I visited the Peruvian archaeology museum in Lima. It took about four million years for mankind to evolve from *Australopithecus* to modern man. I was also able to see the *Homo sapiens* skulls in the museum.

I did not know this until Rumella told me she was going to visit Ethiopia, which was the home of man's earliest ancestors. Until then, I believed that the shore of Lake Turkana in northern Kenya was the place where the ancestor of modern man was found. This was the discovery of Dr Richard Leakey and his team. They discovered a two-million-year-old humanoid skull, which was from *Homo habilis*. In 1974, 'Lucy', a 3.5-million-year-old hominid skeleton was discovered by Dr Donald Johanson at Hardar on the lower Awash River in Ethiopia's Danakil region. Lucy was kept in a museum in Addis Ababa, and Rumella brought me a photograph of Lucy when she returned from Ethiopia in April 2006.

Following human evolution, early migration from Africa started. The Ice Age also played an important role.

Knowledge of human origins and migration is based on the opinion of paleoanthropologists, but now genetics is also able to demonstrate it by DNA studies of the migration pattern. A small group of modern humans left Africa 70,000 to 50,000 years ago. 50,000 years ago, modern humans followed a coastal route along southern Asia and reached Australia. Paleoanthropologists used to believe that modern human migration to Europe took place through the north African route, but genetic data shows that an inland migration from Asia to Europe occurred between 40,000 and 30,000 years ago. It was not from north Africa, as DNA of today's western Eurasians resembles that of India. Around 40,000 years ago, humans pushed into central Asia and some of them travelled north of the Himalayas to south-east Asia and China, and then Siberia and Japan. Between 20,000 and 15,000 years ago, humans from northern Asia migrated to the Americas – genetic evidence supports this. Humans migrated to America when Siberia and Alaska were connected by land, as sea levels at that time were low.

The settled way of life started in about 3000 BC and the rise of human civilisation began. Historians say that the first civilisation was the Sumerian civilisation, which originated in about 3500 BC in Mesopotamia (3500–500 BC). Then came the Egyptian civilisation (3000–500 BC), the Indus Valley civilisation (2500– 500 BC), Chinese civilisation (2200–500 BC), Mediterranean civilisation (2000–500 BC) and the Mesoamerican civilisation (1200–400 BC). Later came the Classical world (550 BC–AD 700), when the civilisations of Greece, Rome, Persia, China and India started to spread. Since those ancient times, there have been civilisations prosperous with a mixture of arts, cultures and the blood of invaders and migrants. The Silk Route played an important part in connecting China, the Middle East and the Mediterranean and their commercial development. The Silk Route can be traced back to the second century BC, but it did not become famous until the Romans showed an interest in silk. The Silk Route became the most important connection between east and west until the discovery of a sea route from Europe to Asia in the late fifteenth century. We saw the starting point of the Silk Route when we visited ancient city of Xi'an, the provincial capital of modern China's Shaanxi Province, in 2004. In 2002, we saw one of the Mediterranean parts of the Silk Road, the thirteenth- century Silk Road through Anatolia, which was the Turkish *kervansaray,* meaning 'caravan place', and a kind of thirteenth-century luxury motel.

It would be interesting if scientists could do a genetic DNA study along the Silk Route from ancient China to Rome, once the capital of the Roman Empire and the cradle of the Mediterranean civilisation, to find out the genetic population migration pattern.

The Crusaders, Mongols, Ottoman Turks, and English and European merchants brought new cultures and workers to various parts of the world.

The Renaissance in Italy, and the age of exploration to the New World, gave a new spin to migration. Individuals started to migrate in search of their fortune, and the slave trade began in the fourteenth and fifteenth centuries from Africa, initially to Europe and then to the New World. The idea of migration and the international labour market started to grow. In 1550, the first slave ship sailed from Africa to the West Indies to meet the need for intensive field labour in the sugar and tobacco plantations.

The slave trade was responsible for the largest mass migrations in human history, and it is now known that about forty million people in America and in the Caribbean are descended from slaves.

Slavery was replaced by indentured or contract labour from about 1830 onwards; workers from India, China and the Pacific went to British, French, German and Dutch colonies in various parts of the world, mainly Africa, Asia, the Pacific, and North and Latin America. Their numbers were thirty-seven million, of which thirty million were from India. After contracts expired, twenty-four million from India returned and the rest stayed. The Indian communities in the Caribbean, East Africa and Fiji are mainly descendants of those workers.

Voluntary mass migration from Europe to America took place in the nineteenth century and, between 1846 and 1890, seventeen million people left Europe; eight million were from the British Isles. Industrialisation in Britain and the potato famine in Ireland (1845–47) were responsible for such migration. The next peak of migration from Europe was between 1891 and 1920, when twenty-seven million people left Europe, the majority from southern and eastern Europe. It was estimated that, over the period 1846–1939, fifty-one million people migrated from Europe. Of these, thirty-eight million went to the USA, seven million to Canada, seven million to Argentina, 4.6 million to Brazil, and 2.5 million to Australia, New Zealand and South Africa.

After the Second World War, there was a massive population movement throughout the world. Fifteen million people were involved in relocation within Europe, and many Europeans migrated to Australia, Canada, the USA, South Africa and Israel. Israel was a newly created state – a home for Jewish people – but Middle East conflicts are ongoing, involving Palestine and Jewish settlers.

The world also witnessed the massive killing and mass migration of population that took place in the Indian subcontinent in 1947, after the partition of British India created two countries, India and Pakistan, following India's independence from British rule.

I was a child, but I still remember the riots and the Calcutta killings in 1946–47. Calcutta was finding it difficult to control mobs and riots. Muslims killed and looted Hindu properties and, in revenge, Hindus and Sikhs killed Muslims. I heard from a doctor who at that time was a medical student; how mobs were killing people, and bodies were lying about on the roads, just 200–300 metres from the medical college. Many were wounded; bleeding and shattered bodies lay all over the road. Red Cross trucks and ambulances were involved in the rescue operation. The doctor, as a young medical student, joined the rescue team but he was horrified by the incident. He was Dr Pranab Barman Ray, who later became a general practitioner in Peterborough, England, where I worked for a short period in his practice.

I remember some of the incidents involving our locality and our family. There was a rumour that a mob was approaching the road where we used to live. Everybody in the house, and our neighbours, were alert with searchlights, guns and binoculars in case of any mishap. Moreover, there was a Sikh family who had a transport business and used to live at the end of the road, and the Sikh gentleman, who was a big man with a sword, used to challenge passers-by if they approached our road. Although our whole family was worried, fortunately there was no catastrophic incident, except a body was found in a manhole on a nearby road.

In 1946, my mother was expecting my sister and, on the early morning of 26 October, she had labour pains. Due to the riots and violence in the city, Calcutta was under curfew. My father was finding it difficult to get hold of our family gynaecologist. My father and other elderly family members were worried and discussing what could be done in the city with the curfew and the riot. However, my father somehow got permission to go out during curfew hours, got hold of our family gynaecologist and was able to bring him to our home. My sister, Menakshi, was delivered at home that afternoon. Although I was a child, I still remember the worries and anxiety that our family underwent in those years.

Bengal was partitioned into two: two-thirds of it joined Pakistan and one-third remained with India. I saw how Calcutta was swelling up with refugees from East Bengal (the newly created East Pakistan). They were all over the place, on the stations and along the railway tracks, and even the abandoned huts in Calcutta Lake Hospital became shelters for refugees. The Calcutta Lake Hospital was no longer operating, but it affected the local rowing clubs. The most important of these clubs is the Calcutta Rowing Club. It is the oldest rowing club in India, established in 1858, the pride of Calcutta since the British Raj. In the year 2008, the club celebrated 150 years of its existence. Prior to celebration, Mr Chandan Roy Chowdhury, the Honorary Secretary, and his wife came to Britain to invite some of the club's oldest surviving members. While they were with us at our home in Hampshire, we were discussing the slums and refugee issue that had affected the club. He told me that, after a long battle, the refugees were moved from the local slum area and resettled to other neighbourhoods and the entire area has developed.

Eventually, some of our relatives who used to live in East Bengal joined us and gradually there were problems of accommodation in our house at Ballygunge in south Calcutta. At the end of 1956, my parents decided to move to the suburbs of Calcutta, where they built a house with a beautiful garden. It was a small village called Baishnabghata and it was under the Tollygunge Municipality. The village was surrounded by various refugee colonies where the refugees from East Bengal settled.

Although it was a nice house, our move in my childhood from Ballygunge to Baishnabghata made me feel like a refugee, and it took quite some time to settle in. This I think might be due to the loss of a city environment, the loss of childhood and school friends, cubs and scouts activities.

To overcome this, we established a club. Apart from sports activities, it had some social, welfare and community services and activities. This gave me the opportunity of working with international organisations such as UNICEF, Oxfam, Christian Aid, Catholic relief services, Ramakrishna Mission, and Bharat Sevasram Sanga. Through the club, we worked for the local refugees and poor people, by organising and distributing food, milk, clothes and blankets, and getting involved in other community programmes or activities.

The Baishnabghata (or Vaishnavghata) is situated on the bank of the Adiganga (Tolly's Nullah), where the main flow of the Bhagirathi to the sea used to be. The course was completely changed and of course it is now deserted as the Old Nullah. Some of the anthropological discoveries along the Adiganga support the theory that one time it was an important sea route.

There is a story of how the name of the village originated. Sri Chaitanya (1485–1534), who was one of the great Vaisnab saints in Bengal Vaishnavism, was travelling along the Adiganga (Bhagirathi) with his disciples from Nabadwip, which was a great centre of Sanskrit learning, situated on the bank of the Ganges, on a pilgrimage to Jaganthdam Puri, Orissa. Some of his disciples stopped on the *ghat* of the village, liked the place and settled. As they were followers of Vaishnavism (Krishna consciousness – worshipping Krishna, the Hindu God), they called the place 'Vaishnavghata' or 'Baishnabghata', and thus the name originated. However, there is no historical evidence to support this, except that the place Baishnabghata is mentioned in old Vaishnava literature.

In 1965, this place became part of the Calcutta Corporation, and my father became the first councillor of this Tollygunge area. I lived in Baishnabghata up until 1967 and, in 1968, moved to the UK. Before I left Baishnabghata, I saw some of the political violence, which did not agree with the democratic political system of the country.

In 1971, Bangladesh was created, which again led to the migration of refugees from former East Pakistan to India. I was in the UK and had the opportunity of visiting some of the refugee camps in India, near the Bangladeshi border. This time, I accompanied some renowned Calcutta-based journalists, including my father and Mr Barun Sengupta. Mr Sengupta was the editor and managing director of a Kolkata-based Bengali newspaper.

After the devastating effects of the Second World War, the reconstruction of Europe took place and there was migration of workers, especially to Germany, France and the UK. The migration pattern was interesting and depended on the nature of recruitment. Initially, recruitment was from displaced people from the war, followed by those from less industrialised European countries like Italy, Portugal and Spain. Germany looked to Turkey and other eastern European countries like Yugoslavia. France and the UK looked to their colonial ties, and thus France recruited from northern Africa and the UK from the Caribbean and the Indian subcontinent.

There were great influxes of refugees from eastern Africa, mostly from the former British colonies. They were originally from the Indian subcontinent, and for generations had lived in Africa; Idi Amin's policies made them leave their country and settle in Britain. They were concentrated in places like Leicester and London, and their contribution to the British economy was great. They kept their own cultures, religions, customs and food habits.

From the 1970s onwards, especially in the 1970s and 1980s, there were chronic shortages of labour in the Middle East, mainly in the oil-rich Arab states, resulting in a migration of labour from Asia.

Patterns of migration are interesting; some are professionals or skilled workers, some are contractual and others are unauthorised workers; some are settlers and others are refugees and asylum seekers. The professionals are mostly experts who move from country to country, and their numbers are small. Contract workers are the workers who are admitted to other countries and allowed to stay for a specific period (that is, on the basis of the length of their contract). Settlers are the people who intend to live permanently in their new country. Unauthorised workers are the people who enter the country and work without any valid documents. They are usually illegal immigrants. A refugee is a person who takes refuge or shelter in a foreign country because of difficulties in his own country, and asylum seekers are those people who have left home and are unable to return for fear of persecution.

The legal definition of such forced migration (Article 1) is:

> *A refugee is a person, who, owing to a well-founded fear of being prosecuted for reasons of race, religion, nationality, membership of a particular social group or political opinion, is outside the country of his nationality and is unable or, owing to such fear, is unwilling to avail himself of the protection of the country; or who, not having*

> *nationality and being outside the country of his formal habitual residence is unable to or, owing to such fear, is unwilling to return to it.*

Environmental refugee or climate refugee is the global warming related environmental disasters such as drought, famine, desertification, flood, melting ice sea level rise, disruption of seasonal weather and so on and the term is generally used when the migration is consider to be forced and not a matter of individual choices.

So, the migration could be voluntary or involuntary. The involuntary migrations are due to slave trade, human trafficking, ethnic cleansing and environmental issues. Whether it is voluntary or involuntary, sometimes people are forced to migrate and some of the factors for forced migrations are:

a) Lack of opportunities and jobs
b) Political fear, partition, prosecution, violence, death threats and war.
c) Loss of wealth, poor housing, landlord /tenets issue, poor medical care and primitive condition
d) Natural disasters including famine, droughts, flood storm, desertification and so on.

The demand for immigrants fluctuates according to economic cycles. During a period of rapid growth, there is a huge demand for workers, and immigrants are brought in to fill the gaps. Immigrants are filling the needs of a wide spectrum of jobs within the labour market. At the top, there are the people with professional skills, such as doctors, engineers and scientists, and at the other extreme, there are the unskilled jobs, the 'dirty, difficult and dangerous' jobs that local people do not want to do.

An example of this involved the Chinese cockle-pickers who drowned in the sea tide at Morecambe Bay, England, while picking cockles. This was a great issue in 2004, as they were illegal migrant workers, engaged in low-paid, dangerous jobs under Chinese gang masters. Sometimes immigrants are also employed in declining industries. Nevertheless, some people protest that immigrants take the jobs of native workers.

In the UK, a large proportion of highly skilled immigrants are doctors and many have been employed in the National Health Service (NHS). There are refugee doctors, too, and the NHS has found that refugee doctors are a valuable resource: according to the British Medical Association, there are 2,000 refugee doctors in the UK.

There are also questions of 'brain drain', which is not a new concept, as young professionals are always looking for new opportunities. Sometimes both the emigrating countries as well as the recipient countries get the benefit of the immigration, although history suggests that some countries get richer by using immigrant workers.

During an economic crisis, there is a halt to the employment of immigrants, and long-term immigrants are more likely to be made unemployed in comparison with native workers. This is because they are employed in more unstable jobs, are vulnerable to discrimination, and so on. In the UK, the unemployment rate shows that 5 per cent are UK-born but 6 per cent are foreign- born. However, a different picture is noticed in France, where unemployment is around 11 per cent and more than 30 per cent are non-EU immigrants.

After 1973, migration became increasingly difficult for people from developing countries who wanted to enter rich countries, especially in Europe. Recently, terrorists and immigration populations are also great issues.

However, some of the rich countries of the world are now facing a population decline. Germany is one of them, and is very concerned about it. The fertility rate in Europe is 1.4 births per woman, but there is also an increase in the elderly population, resulting in a serious impact on pensions. In many countries, immigration has helped population growth, and there is a question of whether immigration can help rejuvenate the population.

In the UK, the population changes all the time but the latest prediction suggest that the UK population is growing; it seems that in 2080, it will reach 84.87 million. There might be a question of the impact on the number as a result of immigration and asylum seekers.

The certain facts and figures, published in the Guardian and Observer (April 2010) are interesting to note. Between mid-2007 and mid-2008, the natural increase of the UK's population was

220,000(adjusted by birth and death). In the same time, due to immigration and other causes, the number was 187,300. Since 1997, approximately 6 million people have entered Britain and 4 million have left. However, with the onset of recession the number has fallen; it was 233,000 people in 2007 down to 163,000 in 2008. Applications for asylum in UK also fluctuated. In 1999, there were 71,155 asylum applicants in contrast to 24,250 in 2009.In fourth quarter of 2009, the eight top countries from where asylum applicants came, were Zimbabwe(5,540), Afghanistan(3,330), Iran (1,835),Eritrea(1,350),Pakistan(1,275), China & Taiwan(1,165), Sri Lanka(1,110) and Somalia (920). The countries and the numbers vary and largely dependent upon the political situation of the seeker's countries at that time. For example, in 2002 Iraq was at the top of the applicants' list but in 2009 it the number of applicants was only 845.

Some people rely on the help of smugglers to enter countries illegally. People are smuggled in by hiding in trucks, by supplying false passports or by bribing immigration officials.

From time to time, there is big news about how illegal immigrants are captured or have died in sealed containers passing through the Channel Tunnel into the UK. Trafficking is another form of migration, in which people are taken by force, or are kidnapped or deceived. Girls and young women are a vulnerable group in particular, taken for sex work or low-paid jobs.

In recent years, family entry and family reunification is a widely accepted policy in most countries. However, sometimes illegal immigrants take advantage of this. There are incidences where false marriages take place to obtain entry into a country and, once the person gets legal status, the couple separates by getting divorced, or one of the partners disappears. In certain communities, there are problems of having more than one wife, resulting in large family migration. It is usually Asian people who bring large families, rather than European or Latin American people. Some developed and rich countries attract or invite foreign students and subsequently allow to them to stay. This is more common at doctorate or post-doctoral levels.

Welfare benefit is a big issue in certain countries, and a common complaint is that immigrants exploit public services. Studies in the

USA in the 1990s concluded that immigrants claimed marginally more welfare than natives, because they are poorer. Low-income immigrants are more likely to claim welfare than low-income natives.

Research in Australia, Sweden and the UK has shown that refugees and asylum seekers often suffer from various mental and physical disorders, which need effective intervention and specialist healthcare provision. Asylum seekers have more problems than ordinary immigrants and refugees, including the issues of health and language. In the UK, to overcome the language problems, demands for ESOL (English Speaker of Other Languages) teachers are growing.

There are certain types of medical condition that are very common amongst immigrants, and sometimes migrants carry some infectious diseases that are prevalent in their country of origin. Some countries demand health checks, immunisation and vaccinations before entering the country and this is certainly true in respect of migrant workers taking up certain jobs in their accepted country.

In the UK, there are certain grievances that, in some local authority areas, immigrants are getting better housing, schooling and hospital facilities, and are more dependent on social service benefits than the native population. It is very difficult to say how much truth is behind this, but there are also propaganda issues from certain native sectors, resulting in growing support for the fascist movement. This has been noticed not only in the UK but also in Europe. There are clashes, tensions and racial violence between immigrant populations and fascist or racist groups. Anti- immigrant politicians are a greater problem in continental Europe than in the UK. This is evident from the continuing support in France for Jean Marie Le Pen, who came second in the 2002 presidential elections, and the success enjoyed by Pim Fortuyn in Holland before his assassination.

In the UK, the housing problem is not a new issue, and at one time in certain areas it was very difficult for immigrants to get appropriate housing or accommodation, as vacancies were denied to ethnic migrants, which was racially motivated. This situation has most probably improved now. However, for this reason, ethnic groups often try to depend upon their own racial group for accommodation and other activities, resulting in closed communities.

Immigrants sometimes try to stay in these closed communities, with their own culture, religion, language, dress and food habits, aloof from the local community and resisting full integration. Sometimes immigrants cannot leave behind their own traditions or habits. This is more noticeable as people get old.

There is a growing tendency towards religious fanaticism that might not be acceptable to the native people. However, certain countries such as Britain want to be a multicultural society, which might or might not be successful.

France has taken the brave step of banning religious clothing in its public schools and the law aims to safeguard the French principle of secularism, but this did not prevent an international outcry from some religious communities. Moreover, woman wearing *burkha* or veil are banned in public places in certain countries in Europe such as Belgium, Italy and France.

However, to create a better relationship, the control of immigration, greater tolerance among the native population and more integration as well as adjustments between the immigrant and native populations would likely improve race relations.

It has been observed that immigrants spend more money on their children's education in comparison with the native population, resulting in a better outcome for second generations.

Not all migration is rosy. Some immigrants return out of disappointment. Many people, who migrate, aiming for long-term settlement, change their minds after spending a couple of years abroad. Some return after the end of a contract. Others leave when they have accumulated sufficient funds or have gained certain academic achievements, and return to their home country with more attractive prospects.

There are also success stories of immigrant populations who have settled. In the UK, the USA, Europe, Canada and Australia, there are migrant populations who are without doubt successful in business, education, politics, science and technology, and other professional activities. They contribute to the development of their adopted country and thus the country becomes richer and more prosperous.

Chapter Fourteen
Global Climate Change

Barrier and wind power, Netherlands.

If one reads through the chapters of my book, one will no doubt find that our global climate is changing. Extreme and frequent storms, hurricanes, unusual levels of rainfall, severe flooding, melting glacier, rising sea-levels and ocean temperature, coral bleaching, El Nino activities, heat wave, drought, forest fire, avalanche, landslides, mudslides, and the changing ecosystem all indicate the extent of global environmental change. Most scientists now belief that these are due to global warming, though many politicians and members of the public do not agree that it is caused by human activities and currently its existence has become a subject for national and international debate.

Since the creation of Earth, the Earth's atmosphere is constant change. This was been gradual and slow but since the industrial revolution the change has been rapid. Prior to the industrial revolution, most change was due to the effects of natural green house, primarily the effects of volcanic eruption where large quantities dust, ash and gas were ejected into the atmosphere. Other factors include continental drift, tectonic shift, comet impact, and so on. However, currently the great area of concern centres on global warming as a result of human activities, which add to the natural greenhouse effect.

The evidence of climate change in the Earth's history supports this. Though it is not easy to investigate the history of climate change, scientists have been able to create methods by which it can be done.

One method is called '*ice cores*' by which samples of ice are drilled and removed from up to a depth of three kilometres in the Arctic and Antarctic. Similarly sediments are taken out from lake, and ocean beds which indicate the amount and type of materials deposited. Past climates can also be investigated by studying tree rings, that is dendrochronology and fossil remains. Satellite images of the Earth from the space with its oceans, deserts, rivers, mountains, lakes and lands help to identify lost or newly emerged land mass from the Earth's past history.

Earth, as a part of the solar system originated about 4,600 billion years ago, in a nebula of dust and gas. About four billion years ago the Earth began to develop a climate with the formation of the continents, oceans and earth's atmosphere. Life began 3.8-3.4 billion years ago.

Seventy to one hundred million years ago, the earth's temperature reached its warmest point with average temperature of 5.C to 12.C warmer than the current earth's temperature. There were no human population, no ice, but high sea levels and a high level of atmospheric carbon dioxide. At that time the dinosaur was the dominant species. Homo Sapiens man appeared 100,000 – 200,000 years ago.

Ice ages have been a regular occurrence in the Earth's history. The last major ice age was 18,000 to 20,000 years ago and the Earth's temperature was -6.C as thirty percent of the earth's surface was covered with ice. The sea level was low and more lands were exposed; many islands were joined together. The human population was low and the atmospheric levels of carbon dioxide were not high. Scientists also belief that there was a little ice age during the period between 1550 and 1850, when some of the glaciers expended.

Since the eighteenth century the climate has changed due to industrial activities as well as burning of fossil fuels by human. Currently 6.7 billion people exist on Earth and they are producing more and more greenhouse gasses which accumulate in the atmosphere. Global greenhouse gas emissions are mainly from industry, energy supplies and transport, though the level of gases can rise and fall as a result of natural process too. Transport accounts for about fourteen percent of global greenhouse gas emissions and road traffic is the main factor. Energy uses in buildings are the biggest factor in greenhouse emission. A recent British report revealed that twenty-seven percent is from domestic use and eighteen percent is from offices, factories, distribution depots, schools, stores and shops, in total, accounting for nearly 266 million tones of carbon dioxide which was released in the atmosphere in 2008.

Over the past one thousand years carbon emissions have risen above 8 Giga-tons of carbon per year; carbon dioxide concentration was more then 376 parts per million and the temperature has risen above 0.6.C. At present 10.4 per cent of the Earth is covered with ice and there has been a forty per cent increase in atmospheric carbon dioxide since the commencement of the industrial activities. According to the World Resources Institute (EIA, Times research),total carbon dioxide emissions by individual countries in 2007 included: China 6,284 million metric tonnes; USA 6,007 million metric tones; European Union 4,267 million metric tonnes; Russia 1,673 million metric tonnes, India 1,401 million metric tonnes; Brazil 398 million metric tonnes. Cumulative world wide carbon emissions from 1850 to 2002, showed that the USA produced 29.3 percent followed by the European Union 26.5 percent, Russia 8.1 percent, China 7.6 percent, India 2.2 percent and Brazil 0.8 percent.

The first direct measurements of atmospheric carbon dioxide concentration started in 1958 in Hawaii, USA. This is the location of the high altitude atmospheric research station, situated at the submit of Mauna Loa, at an altitude of about 4,000 meters, which I sited when I flew over by helicopter in March 2008. There are dome houses which accommodate solar radiation instruments and that measure the effects of atmospheric dust on sunlight. The observatory monitors all atmospheric constituents such as greenhouse gases and aerosols that may contribute to the climate change.

Besides carbon dioxide and carbon, other greenhouse gases include water vapour, methane, halocarbons, and nitrous oxide. Water accounts for sixty percent of the natural greenhouse effect. According to a recent report methane emissions have increased by 31 percent in the Arctic over last five years because of melting frozen Arctic soils. Other methane emissions are from the Amazon basin, Congo basin and paddy fields in Asia. Landfills and farming are the human factors responsible for methane emission. Halocarbons including chlorofluorocarbons (CFCs), used in refrigeration, insulation and spray–can propellants reacts with ozone, resulting in ozone depletion in the earth's atmosphere. Ozone absorbs much of the incoming solar ultraviolet radiation (UVR) and the destruction of ozone causes more ultraviolet radiation, which is damaging to human health. There has been a fifteen percent increase in levels of nitrous oxide since the pre-industrial era. This is primarily due to industry and agriculture. Nitrous oxide is released when farmers use nitrogen-based fertilizers.

Burning fossil fuels (oil, coal, and gas) is the main contributor to polluting the atmosphere and, according to one estimation, three million people die annually from the effects of the pollutants such as sulphur dioxide, nitrogen dioxide, particulates, ozone and others. Sulphur dioxide is the industrial pollutant that forms the acid rain. Nitrogen dioxide comes from vehicle emission that causes smog. Particulates create dust and smoke that affect the quality of air.

Two-thirds of our planet is covered by oceans which are increasingly at risk as large numbers of pollutants enter the seas. These include sewage, farm run-off, air pollutants, maritime transportation, industrial wastewater, offshore oil and litter. It is said that about sixty percent of coral reefs and marine products are threatened by human actions. Reefs are built of from coral skeletons and held together by layers of calcium carbonate produced by coralline algae. The coral reefs are under treat from climate change, ocean acidification, overuse of the reefs resources and harmful land use practices.

About 97.5 percent of the Earth's water is salt water and the remaining 2.5 percent is fresh water. Fresh water is comprised of 68.7 percent glaciers water, 30.1 percent ground water, 0.8 percent permafrost and the rest 0.4 percent surface and atmospheric water (freshwater lakes 67.4 percent; soil moisture 12.2 percent; atmosphere 9.5 percent; wetlands 8.5 percent; rivers 1.6 percent; and biots 0.8 percent). The demand for water is mainly for agriculture use, industrial and domestic use. Millions of people do not have access to clean drinking water and the demand for water will only increase as the world population grows. Lack of rain, drought, raised sea water levels, salinity as a result of climate change, all create more problems regarding access to water.

Recently, I travelled to the Ganges delta and found that level of ground water has changed and that now, in order to extract ground water, one has to bore deeper into the Earth's soil. This is due to over withdrawal of ground water by the ever increasing population. Moreover, the quality of water has been changed and there is now more salinity in the water. In addition a local scientist stated that there are some ecological changes as mangrove plants can now be found growing along the Ganges round the Kolkata which is unusual.

All over the Earth, greater soil degradation and erosion is happening as a result of human activity such as overgrazing, deforestation, agriculture, wood gathering for fuel and industrial practices, which have a great impact on the Earth's climate. Most deforestation occurs in South-east Asia, South America and Africa. Forests capture large amounts of carbon dioxide, releasing oxygen and protect the soil, fresh water, local habitats and various species. The loss of forests will mean the break down of this circle. Coal-fired power stations and vehicles are the major emitters of carbon dioxide, the main factor of man-made climate change. Global climate change will have major effects on health which are already occurring. More people will suffer from coughs, breathing issues, chest pain, respiratory tract infections, eye diseases, rhino-sinusitis and lung diseases. People with hay fever, asthma, and chronic obstructive lung diseases will get worse. Stomach related gastro-intestinal diseases and parasitic diseases will increase.

There will be more cases of certain cancers including skin and lung cancer. There might be an effect on immunity and the activation of infection. The re-introduction of malaria in certain areas and spread of tick-bone disease, lyme disease will occur. Extremes of heat and cold can cause fatal illness. Chronic food and water insecurity will have an indirect impact on health.

Scientists have already noticed a number of events occurring due to warming of the earth's atmosphere. Glaciers and ice sheets are melting. Mountain glaciers are thinning or reducing in size. The glaciers in the Alps have been sinking so rapidly that Italy and Switzerland have had to redraw their national borders. The Glaciers of the Himalayas are receding and this will have a great impact on water flow as the major rivers of South Asia originate from the lakes and glaciers of Himalayan. As a result the river flow might have to rely on monsoon water rather then glacier water, although reliability of the certain data on glaciers is questionable. Mount Kilimanjaro in Tanzania has lost more then eighty percent of the size of its glacier since 1912. The Glacier National Park in the USA had 150 glaciers in twentieth century and by 2005 it was only twenty-seven. The Quelccaya ice cap in Peru is loosing its edges every year. Many ice shelves in Antarctica have been receding or collapsing and the great Greenland ice sheet is loosing its ice at a rate of 239 cubic kilometres per year. The sea level is rising, so is the ocean temperature.

About four-fifths of Greenland is covered with an inland ice sheet that is located centrally in an area of 1,710,000 square kilometres (660,235 square miles). Each year about 600 cubic kilometres of ice accumulates on the ice sheet and a similar amount is lost through melting and by the production of icebergs that are calved out from the glacier and flow into the sea. Currently, rising temperatures have led to an increasing loss of mass from the ice sheet and a subsequent increase in global sea levels. In the year 2007 the Greenland ice sheet lost 592 cubic kilometres of its mass. It is melting at a rate that is 30 per cent higher than 40 years ago. Is the loss in mass due more to the ice flow from the outlet glacier than to the melting of the ice sheet?

Many scientists think that the increased loss has been due to a roughly equal increase in the speed of the glacier flowing into the ocean and increased melting at the surface. Current investigations by scientists from British universities along with the world's scientific organisations suggest that the water is flowing all the way to the bottom of the glacier through a vast internal plumbing system that acts as a lubricant between the ice and the rock underneath. This jacks up the icesheet, allowing it to slide downhill, over the rock and into the ocean.

It is not easy to find out the total number of Greenland glaciers; however it seems that there at least 36 major glaciers of which many are large outlet glaciers. Most of the drainage takes place on the western rather than the eastern and southern parts of the country, which are largely covered with high mountains. Along the west coast there are 35 areas of glaciers where minimum to maximum ice production take place; of the 35 areas, the Sermeq Kujalleq (Ilulissat Glacier) is the fastest and most productive glacier in Greenland. It drains 7 per cent of the entire Greenland ice sheet and produces one-tenth of the total production of icebergs from the inland ice. The ice steam discharge reveals that in 1996 it discharged 24 square kilometres per year and this was responsible for a rise of 0.06 millimetres in sea level. The 2005 statistics showed that the discharge increased to 46 square kilometres per year, which was again responsible for a 0.12 mm rise in sea level each year. Since 1998 the ice velocity has increased to 15 kilometres per year from 7 kilometres per year.

The calving ice front of the Ilulissat Glacier has retreated more than 40 kilometres since 1850 and the glacier surface is thinning by 10 metres each year. In 2004 it became a World Heritage site. With these facts and information, I decided to visit this part of Greenland. Moreover, the ocean geography in that part is very interesting. More than 40 million tons of ice flow out into the fjord every day, resulting in large icebergs that escape across the mouth of the icefjord and then drift out through the Disko Bay to the open sea.

Icebergs from the mouth of the Kangia (Ilulissat) drift either south or north. The majority drift towards the north around Disko Island and there enter the Davis Strait between Greenland and Canada. Further drifting depends upon the way in which the various ocean currents operate, as icebergs are usually caught by the West Greenland current, the Baffin and Labrador currents. Some of these icebergs are dangerous to shipping as they enter the North Atlantic shipping route. The sinking of the Titanic in 1912, which took place off the south cost of Newfoundland, might have been due to one of these icebergs.

The Disko Island is famous for the Arctic station research centre which is located one kilometre east of Qeqertarsuaq town. The Arctic station was founded by the Danish botanist Morten Petersen Porsid in 1906 and is within the Faculty of Science of the University of Copenhagen. The Arctic station operates two climate stations that register air temperature, relative humidity, wind speed, ultra violet radiation, and so on.

To observe the effects of climate change, I visited the Disko Bay region in August 2010, which included the Ilulissat and Arctic stations. To get there, I flew from Copenhagen to Greenland International airport in Kangerlussuaq. Kangerlussuaq airport is a small airport and I had a three hour wait before I could board the domestic flight to my next destination, a place called Aasiaat. I came out of the airport lounge and found that many passengers were sitting in the sun. It was a lovely day and the temperature was warm enough to enjoy the mid-day sun. However I noticed that the sun radiation was so powerful that one could very easily tan within a very short period.

In 2010 Greenland is more than 5°C (9°F) warmer than the average between 1950 and 1980. This increased temperature is no doubt one of the reasons for the melting of the Greenland ice.

From Kangerlussuaq airport I flew to Aasiaat, and then on a three and a half hour's boat journey through Disko Bay until I reached Qeqertarsuaq town. I had my first glimpse of the icebergs as we were approaching nearer to Qeqertarsuaq or Disko Island.

From a distance, it looked to me like we were going to pass by a couple of white chalk islands that were shining in the sun. However, when our boat passed nearby, I realised that they were floating icebergs. The numbers of icebergs gradually increased as our boat approached Qeqertarsuaq town.

The town is built on volcanic rocks, surrounded by table mountains. The rocks are mostly basalt rocks, formed as a result of the lava flow during the early Tertiary era. I noted further geological features as I walked along the track, east of the Arctic station. I crossed a fast flowing river, called Rode Elv (Red River). The water is formed from the red coloured suspended sediments which had disintegrated from the red volcanic rocks. When I walked south of the Arctic station along the sandy coast, I saw black coloured sands, which were again basalt. I also collected some basalt rocks of which one had white crystals.

The landscape of Greenland consists in general of ice and barren rocks but on Disko Island, I saw some vegetation including flowering plants. According to one source, approximately 250 of a total of 500 flowering plants are to be found there. I was really surprised to see some wild mushrooms growing on the hiking pathway. I wondered whether these mushrooms were the poisonous or edible type?

Later I asked Outi Maria Tervo, a PhD student and scientific leader at the Arctic station. She said that, 'No mushrooms grow here which are poisonous and so they are safe to eat.' However I will be cautious in making such a statement as my medical knowledge says that some variety of mushrooms are harmful to the health; even in Britain there are wild mushrooms that contain more than 20 toxins.

Most of the food in Greenland comes from outside and the local supermarket shelves are stocked full of such products. I was surprised to see that even though there might be some potential for cultivation or to grow vegetables in the summer I did not see any such activities in Disko Island. However, I saw moss growing along the rocks, giving the appearance of green velvet. This type of rocky area is always moist because of the numbers of underground springs. I gather that there are at least 200 springs and some of them are radioactive. A few are also located to the north of the Arctic Station which I went to explore. I have also seen a stream of water flowing down the hill as a small waterfall; this water was collected and then pumped to supply the town through visible pipes.

North of the Arctic station, at a height of 800 metres (2,600 feet), there lies the Lyngmarksbraeen glacier where I was told that in summer time one can drive a dogsled, though when I talked to the person who operates this in late August, he told me they were not able to manage it. He also informed me that the Lyngmarksbraeen glacier is retreating and its ice sheet is thinning out but I was not able to obtain any data on this.

When I was at the Arctic station, Outi Maria Tervo was kind enough to show me round. I was interested in the climate data that was not available there and so I contacted Professor Bo Elberling at the Department of Geography and Geology, Copenhagen University.

However, Outi Maria was able to reveal the interesting information that, although sea levels are rising, Greenland is also rising due to land uplift and she showed the instrument outside the station that is recording these changes. Of course, the climatic history supports the fact that when an ice sheet disappears, the land rises. It has happened in many places in Europe and North America.

After spending 2 days in Queqertarsuaq, I left for Ilulissat. The boat journey began in late afternoon, taking 6 hours by boat from Quqertarsuaq to cross the Disko Bay and to reach Ilulissat. As the boat was approaching Ilulissat town, I saw more and more icebergs, which were also getting larger and larger.

The Ilulissat town is located at the mouth of the icefjord, which is filled up with huge icebergs from Sermeq kujalleq (Ilulissat Glacier). I stayed 3 days in Ilulissat.

On the first day, I went to see the ice-calving glacier, Eqi (Equip Sermia), which is located 80 kilometres north of Ilulissat. It was a whole day trip and the boat was modern and was easily manoeuvred through the ice-filled water. There was a fantastic view as our boat approached the front of the glacier. The boat went as close as possible and I saw the ice-calving phenomenon; masses of ice was breaking away from the glacier and crashing into the water, preceded by loud sounds.

On the boat I could feel the wakes, which were the result of the entry of the ice into water. I was told that sometimes the wakes were so large that the boat could not approach closer as it was often hazardous. The glacier itself is 700 metres wide and the local guide pointed out the areas on both sides of the front of the glacier where nothing but bare rocks could be seen: in the fifteenth century and again in 1910, the ice mass on those areas disappeared or melted away.

Over the next couple of days, I tried to explore Kangia Icefjord (Ilulissat glacier) which is 40 kilometres long and up to 7 kilometres wide with an average depth of 1,100 metres. The two big ice sheets of Sermeq Kangia merge into one called the Sermeq Kujalleq (Ilulissat glacier), which is the most important ice stream in the Greenland. This is the most productive and fastest glacier with an extremely high velocity of

22 kilometres and a production capacity of 20 million tons of ice per day so that huge icebergs float easily towards the open sea beyond the mouth of the icefjord at a sea depth of 200 to 300 metres.

I visited the Ilulissat museum where I saw a depiction of how, year after year, the new glacier front is created after the breaking up of the floating front. The constant changes took place over the period between 1850 and August 2006. In modern days, the mapping of these changes is usually done by satellite monitoring and or by aerial photography, helicopter- bone radar technology, and climate observation.

The Ilulissat glacier has further retreated since 2006. In order to observe this, I took a helicopter ride. The helicopter took off from Ilulissat airport and flew low over hills, lake, icefjords, landing on a preserved area of the mountain at Kangia near the glacier. The helicopter used to land further north and west of Kangia but due to the effect of the retreating glacier the helicopter landing site has been shifted to the south, which is the present site.

After getting down from the helicopter, within walking distance I was able to see the glacier and its breaking front. At the edge of Kangia there were climate instruments that constantly recorded the changes. The helicopter also flew over the glacier, and its edge and break up points. The Ilulissat glacier looked like castles of ice with towers and spires; on the ride back I saw steam from the changing ice drift over the Kangia up to the sea front, in the form of packed ice, thawing ice, floating icebergs and the standard icebergs at the mouth of the icefjord. It was a clear, sunny day and I took aerial photographs of the changes in Sermeq Kujalleq and Kangia. One of the best ones showed the retreating glacier and its break up point; it is printed at the end of this chapter.

Besides the helicopter tour, I also sailed by boat round the giant icebergs which stood at the mouth of Ilulissat Kangia. Some of the icebergs were more than 100 metres above the surface of the water.

However, in 2010, an extraordinarily huge block of ice (260 square kilometres) broke off. It was not from Ilulissat but from the Petermann glacier located in the north-west of Greenland, which indicates that the melting of Arctic is accelerating.

At the end of my visit, my impression was that Greenland is losing its ice very fast. Due to Arctic shrinkage as a result of climate change, the north-west passage, which was inaccessible until 2009, has now opened up. It seems that this will be the new sea route in the Arctic Ocean, which connects the Atlantic and Pacific Oceans.

If global warming continues like this, then low-lying islands, deltas and the coastal areas carry the greatest risk of submerging. The most vulnerable low-lying deltas of the world are the Rhine-Maas-Scheldt in Netherlands, the Nile in Egypt, the Ganges-Brahmaputra delta in Bangladesh, the Mississippi in the USA, and the Yangtze in China. Some of these deltas are thickly populated.

The Netherlands is sinking in relation to sea level at a rate of twenty centimetres per one hundred years. However, if the temperature of the Earth increases faster due to natural and human activities then there is an increased risk of the low-lying area of the Netherlands becoming submerged due to rising sea level as a result of melting polar icecaps.

The last major flood disaster in Netherlands took place on 1 February 1953 when 1,835 people drowned, 200,000 livestock died, and over 47,000 buildings were damaged. Many towns, villages and farmlands were flooded. Sixty-seven dykes and 400 beaches were destroyed. Over the years the Netherlands has lost a lot of lands: on many occasions the coastal defences could not withstand the sea water which over-powered human efforts until the Netherlands developed the delta project, which began in 1958 and was completed in 1997.

The project involved building dams, dykes and storm barriers. Dams such as the Veerse Gat dam, the Haringvilet dam, the Brouwers dam, and the Eastern Scheldt dam were constructed on the Netherlands' four tidal inlets, in the years 1961, 1971, 1971 and 1986 respectively. The sluice complex of the Haringvilet dam also controls the discharge of excess water from the Rhine and Mass into the North Sea. Two inland dams, Zandkreek and Grevelingten, were built in 1960 and in 1965 for the uncontrollable currents in the tidal area. Another inland dam is the Volkerak dam, which was built in 1969 to separate fresh water from salt water. The Phillips dam was built in 1987 and its lock complex is also used for the salt/fresh water separation system. The Oester dam, which was built in 1986, is the longest dam in the delta. (nearly 11 kilometres). The dam is situated on the inlet where there are tidal and non-tidal issues for the shipping route; it was built in the area where Dutch people had always fought against the potentially disastrous power of the sea. During the construction of the Oester dam, strong currents were building up and in order to prevent these, the Markeiezaat dyke was constructed and completed in 1983. The Phillips and Oester dams are the world's largest sand dams, constructed in flowing water using its sand. The Bath Discharge Canal was built in 1987 to discharge excess water from the rivers and polders. The sluice gates and locks were also built to prevent salt water from getting in when the water level is high. The first storm surge barrier was built in the year 1958 on Hollandse Ijssel to protect the low-lying, densely populated region of the Southern Netherlands; the last one was built in 1997, the Maeslant storm surge barrier, which protects Rotterdam and its surrounding area from flooding.

Along with dams and locks, more new roads were constructed, resulting in better road communication, lakes, residential and recreational areas were also created.

In April 2010, we stayed in one such delta area, called Noordzee Residence De Banjaard, which is located on the North Sea shore of the island of Noord-beveland. The place is situated in between the Eastern Scheldt dam and Veerse Gat dam. The Eastern Scheldt dam is on the mouth of the Eastern Scheldt and the Veerse Gat dam is on the North Sea inlet where the lake Veerse Meer was created as part of the delta project.

As part of our journey, my wife and I flew from London Gatwick to Dusseldorf where Salil Chandra and his daughter Nicole received us. We were very pleased to see Salil who, despite suffering from a serious illness, came to the airport to meet us. Unfortunately after a long struggle with cancer, Salil died on 31st July 2010. From there we went to Salil and Marianne's house at Voerd, 60 kilometres from Dusseldorf airport.

The next day they took us to Banjaard, a 280 kilometre drive, which Nicole drove very well. We passed by the Rhine, the Waal and the Maas rivers, along the Netherland's low-lying delta region, arriving at the park area in the afternoon. We stayed in a rented self -catering house in the park for three nights, surrounded by the canal. I was pleased to see evidence of the Holland's environmental awareness in the way that the country is generating electricity by utilising coastal winds. So our landscape was no doubt a picture-square residential park with a canal, protected by the sea dyke, onshore wind turbines and white sandy beaches. We visited local villages, towns, Neeltje Jan's wetlands and explored the various delta project areas, finding out how the Netherlands is preserving the environment and securing the land against flooding.

There is no doubt that the delta project has been a great achievement for Dutch civil and hydraulic engineering; the project was so successful that other countries of the world are researching the same Dutch delta solution.

The Nile delta starts north of Cairo where the world's greatest river bifurcates into two main branches: the Rosetta and Damietta. In between, many small rivers, canals, irrigation ditches, wells and pumps operate so that water is channelled out onto the delta plain before it enters into the Mediterranean Sea. The delta plain is V-shaped a fan delta with more than 10,000 kilometres wide on the coast; 50 million people live in this region. Some scientists believe that the high Aswan dam is responsible for the sinking of the current Nile delta, creating a new delta in the dessert. There is also coastal erosion.

Currently 30 percent of the land mass of the delta is less than a metre above sea level and, in some of the coastal area, it is sinking at a rate of nearly one centimetre per year. Some scientists have already predicted that the sea levels in the Mediterranean are expected to rise by one metre over the next forty years due to global warming and this might cause the loss of one-third of the delta land mass. Many people think that Egypt needs a great delta work like the Dutch but the question is 'Are they able to afford it? 'Another idea is to stop the sea level rising even perhaps by blocking the Mediterranean at Gibraltar. However, the Egyptian government with the United Nations Development Programme (UNDP) is developing a project as part of a climate change adaptation in the low-lying delta area; already a sixteen million dollar project has been launched which includes the strengthening of the sand dune systems, beach improvement and established of the wetlands.

Desertification is another issue. The Sahara is the world's biggest dessert: when we visited part of the dessert in the year 2002, I noted that this desert area was expanding. Examining the history of Earth, one will find that Sahara was once an area with a tropical climate which turned into dessert. This might be due to natural phenomenon but human activities like overgrazing, poor agriculture technique, cutting down the trees for firewood as well as drought are responsible for the current expansions of Sahara deserts.

Many civilisations have been lost in the past due to climate change. The classical example is Petra in Jordan which I had the chance to a visit in February 2010. Now Petra is dry, dusty and ruined city. The first Nabatean and then Roman city is now covered with the dessert sands. This was once a rich city with a good water supply and a flourishing agriculture. We saw the remains of water channels, runnels cut on the rocks and remains of ceramic pipe that once supplied to the famous city of the Petra.

The archaeological record suggests that the Petra climate was once Mediterranean but it is now dry (arid or semi-arid). Many believe that this climate change might have been due to a prolonged and increasing drought that dried up its water system and supply to the city. Recurrent earthquakes did not help either. The city was abandoned. Similarly many cities and lands of old civilisations such as the Egyptians, Sumerians, Indus Valley, Khymer and Chinese were lost due to regional climate change.

The over-extraction of water is a great issue as part of global warning. River, lakes, and ground water are all under constant threat.

History tells us how human activities in the form of over-extraction affected the Aral Sea. In the future, a similar problem with the Dead Sea can not be ruled out. To witness that, I travelled to the Dead Sea in February 2010. The Dead Sea is fifty-five kilometres south-east of Amman, the capital of Jordan and is the lowest point of the Earth's surface, lying at a depth of 1,312 feet (400 metres) below sea level in between the border of Jordan and Israel. It is the part of the Jordan rift valley, which is the deepest hyper saline lake in the world. The salinity is so high that it stings if sea water goes into the eyes. It happened to me once but I recovered quickly after bathing my eyes with pure water. Its salinity is 33.7 percent which is 8.6 times more salty than other seawater, making it very buoyant. I noticed this when I tried to swim in the Dead Sea; the only way I could was by floating on my back, unable to sink except once when I became trapped by the Dead Sea's quick sand or mud. The Dead Sea mud is black and contains various minerals which I learnt are good for the skin, blood circulation, arthritis, and rheumatism.

So, while on the Dead Sea, I saw that many people were applying the black mud to their skin which I also did. To get the black mud, I tried to dig it from the sea floor, away from the gravel area, by hand. I stepped into the quick sand or mud with my right leg which suddenly started to sink. I tried to force my right leg out but the pull was so high that I could not. It was good that I was wearing a pair of rubber slippers and eventually I rescued myself by letting the slipper go from my right foot.

The Dead Sea is fed by the river Jordan. There is no outflow and the atmospheric high temperature and low humidity are responsible for exceptionally high rates of evaporation and the production of large quantities of raw chemicals. These chemicals are used in medicine, agriculture and industries. There are thirty-five different types of minerals and the composition varies with temperature, depth and the season, though they are mainly magnesium, calcium, potassium, bromine, sulphur, sodium and iodine.

The average rainfall is less than 100 millimetres (3.94inches).The maximum summer temperature is 40.C (104.F), though sometimes it goes up to 50.C while its maximum winter temperature is 36.C (86.F). The Dead Sea is rapidly sinking at the rate of one metre (three feet) per year. The water inflow to the Dead Sea from the river Jordan was reduced as a result of large scale irrigation as well as low rainfall. The southern end is fed by a canal, maintained by a company which converts the Dead Sea's raw materials into commercial products.

During my visit I also saw that plenty of agricultural activities were ongoing, which relied on the water supply to the fields. To prevent the Dead Sea sinking many suggested taking action on sources of pollution such as the industrial activities around the Dead Sea. Recently Jordan has been working on the Jordan National Red Sea Development Project

(JRSP): the plan is to extract sea water (about 300 million cubic meters of water per annum) from the Red Sea near Aquaba and convey this water to the Dead Sea. The sea water is to be desalinated along to the route to provide fresh water and the waste water will be supplied to the Dead Sea through a tunnel.

If the Earth's atmosphere changes through global warming then semi-arid regions will become arid, temperate region will develop into a Mediterranean climate, the polar climate will become a cold climate, a cold climate will become temperate, and so on. These will also alter the regional ecosystem. Over the past few years I have noticed some interesting phenomena in our garden. When daffodils dance and birds sing, I certainly know that spring has come. This year it happened in March which is the right time for spring flowers to bloom, but for the past few years, it was in January and this indicates an early spring because of a short and mild winter. The hot, prolonged summer and mild, short winter supports the idea that southern England is developing a Mediterranean climate.

However, this phenomenon might not be so drastic, if there is reduction in solar activities resulting in colder winter, less summer heat and reliable rainfall. Already over past two years there have been colder winters and heavy snowfall in Britain, Europe and North America. Due to Arctic wind, there was a great freeze in Britain recently and this was not typical of the last thirty years. Some areas might benefit from global warming, but overall the risk of human misery is great.

Most countries have now realised, that there will be more variable climate in future as the world continues to warm due to greenhouse gases. To solve some of these problems, the man-made activities responsible for the rising levels of atmospheric carbon dioxide and carbon emissions that are elevating the earth's temperature must be curtailed. The world community has to agree and look for alternative, renewable energy and low carbon emission technology. Other measures include halting deforestation and planting trees and recycling waste water. These all cost money and many countries are not in agreement for economic reasons.

Other factors include change of life style, employment issues and effects on standard of living. Some of the developing countries have given priority to economic developments as it is not easy to compromise on man-made greenhouse activities.

Since 1979, there have been many international conferences on climate change where scientists have urged the world's annual emissions of carbon dioxide should be cut by one-two percent. The last one was United Nation-sponsored Climate change conference at which 192 countries participated; it was held at Bella Centre, Copenhagen, Denmark between 7 and 18 December, 2009.Most nations recognise the scientific view that the increase in global temperature should not be 2 degree Celsius.

To achieve such a target, there should not be any attempts to hamper the economic progress of developing world and a fund will be established to help poorer nation adopt to the threat of climate change with initial outlay of US$30 billion, rising to US$100 billion in 2020.There were many disagreements and the outcome was not very satisfactory. However, on the last day there was some meaningful agreement by USA, China, India, Brazil and South Africa on a target of two degrees Celsius.

Climate change or global warming is complex issue. More research is needed on the health effects of climate change including risk assessments.

A study in 2004 revealed that the NHS in England is responsible for three percent of the UK's total carbon emissions; since April 2010, all NHS Trusts are subject to the government's carbon reduction commitment which demands more than eighty percent reduction of carbon emission in the health service over the next three decades. The question is, 'how will this be implemented?' Each speciality has to take its own lead. According to Frances Mortimer, medical director at the Campaign for Greater Health, the green nephrology project is the first in this kind of initiative. The renal unit initiative includes dialysis water recycling, heat exchanges, reduction of packaging and so on. The green nephrology fellowship is funded by NHS kidney care and managed in partnership with the Renal Association, the British Renal Society and the Campaign for Greener Health care.

The World Medical Association, the Climate and Health Council (UK), the Canadian Association of Physicians for the Environment, the International Society of Doctors for the Environment, Health and Environment Alliance (Europe), Health Planet, Greener Heath Care, Medat, Sustainable Development Unit, and Doctors for the Environment (Australia), are some of the many world organisations who are campaigning for climate change and good health.

Many have signed up to lead a new low carbon lifestyle and also to opt for a low carbon health service. 5,863 health professionals have signed: 115 countries were involved. However, the latest information I have received from Mike Gill and Robin Stott of the climate and Health Council is that now there are 6,000 signatures pledged from 112 countries. In the UK, the climate & Health Council was formed in the year 2006/2007 by leading medical and healthcare professionals including the editors of the Lancet and the British Medical Journal, the President of the Royal College of Physicians, the President of the Faculty of Public Health, and the President of the Standing Committee of European doctors. Now they have formed an international group.

It is good that many doctors, health professionals and other concerned colleagues are already working together to promote world environmental health with a regional emphasis on climate change.

It seems that climate change and health protection is the common goal.

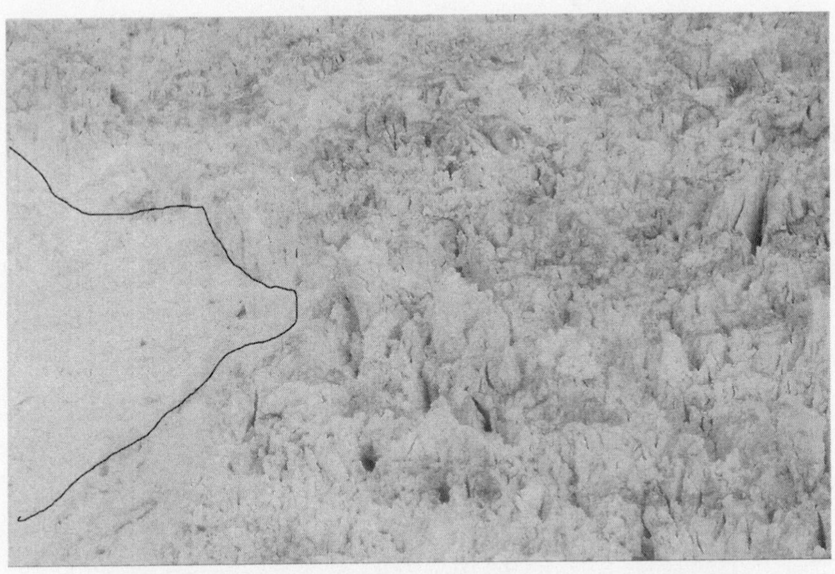

Retreated ice front, Sermeq Kujalleq Glacier,Greenland-2010

Index

D

G

M

N

O

S